PROCUREMENT IN THE AGE OF AI

A PRACTITIONER'S GUIDE TO TRANSFORMING HOW YOU SOURCE, NEGOTIATE, AND CREATE VALUE

DAVID JOHNSON

Procurement In The Age Of AI

CONTENTS

CONTENTS

FOREWORD

BY BRAD VEECH

For decades, procurement has frequently operated as a corporate blind spot. The function that determines who an organization buys from, how much it pays, and how it mitigates operational risk has too often been managed as a back-office administrative necessity rather than a front-line strategic weapon.

David Johnson has spent thirty years proving that it is exactly that, a strategic weapon. He has sat across complex negotiation tables, built supplier partnerships that weathered unpredicted global crises, and taught the next generation of professionals at Rutgers. The insights in these pages are not academic theory; they are earned in trenches.

What David has structured here balances two distinct but dependent concepts.

First, it is a masterclass in the enduring core principles of great sourcing: the total cost of ownership over short-sighted price obsession, the compound value of genuine supplier collaboration, and the intense preparation required to create real commercial value. These fundamentals have not changed through the years, and they will not change in the future.

Second, it is one of the most grounded, actionable accounts of what artificial intelligence is actually doing in modern procurement. David successfully resists both breathless hype and skepticism. Instead, he delivers absolute clarity. What technology can do, what it cannot do, and how a practitioner can leverage it.

That practical specificity is what makes this book highly valuable. The supplier intelligence workflows and negotiation preparation frameworks are not concepts to admire from a distance; they are step-by-step checklists to execute.

Two arguments within this book deserve particular attention:

- **The Reversal of Information Asymmetry:** Historically, sophisticated enterprise sellers held a structural data advantage over buyers. David demonstrates how AI is flipping this dynamic. A well-prepared procurement professional who knows how to direct these tools can now uncover market intelligence, cost structures, and competitive positions with unprecedented speed, erasing the traditional seller edge.
- **Human Amplification over Replacement:** Rather than a narrative of job obsolescence, David makes a compelling case for an inverted workflow. By allowing AI to handle

> the heavy lifting of data collection, classification, and analysis, category managers can shift their time from administrative formatting to high-value strategic execution, relationship building, and critical judgment.

Practitioners reading this book face a moment of genuine optionality. The capabilities described here are available now, and the gap between AI-augmented procurement and traditional methodologies will only continue to compound.

The best procurement executives have always been, at their core, intelligent and strategic professionals who understand that the quality of their agreements depends entirely on the quality of their information. David Johnson has taken that specialized intelligence capability and made it accessible to practitioners at every stage of their careers, not by shortcutting human expertise, but by amplifying it.

Review these workflows, apply the protocols to your next negotiation, and keep this guide close at hand as the profession continues to evolve.

PART ONE
THE FOUNDATION

THE PRINCIPLES, BEHAVIORS, AND CRAFT THAT MAKE PROCUREMENT MATTER—AND THAT AI DOES NOT REPLACE.

HOW WE GOT HERE:

WHY PROCUREMENT MATTERS MORE THAN YOU THINK

Imagine you wake up one morning, you're craving fried chicken. Not just any fried chicken (you want the kind with eleven herbs and spices, the kind that's been around since 1952, the kind with the red-and-white bucket that practically screams "I'm not cooking tonight." You grab your keys, drive to your nearest KFC, walk through that door full of hope, full of hunger, and full of anticipation) and the person behind the counter looks at you with the saddest eyes you've ever seen and says: "I'm sorry. We don't have any chicken."

Read that again.

A chicken restaurant. No chicken.

This isn't some hypothetical cooked up in a business school case study. This actually happened. In February 2018, KFC UK ran out of chicken at nearly nine hundred locations across the United Kingdom. Nine hundred restaurants. Shuttered. Locked. Dark. Because there was no chicken. Lines of bewildered British citizens stood outside

KFC locations like they were waiting for a museum that had forgotten to display its exhibits. The internet had a field day. The hashtag #KFCCrisis trended across social media. People called emergency hotlines, actual emergency hotlines, to complain. The UK government issued a statement gently reminding citizens that the absence of fried chicken did not, in fact, constitute a national emergency.

So how does a chicken restaurant run out of chicken?

The answer is procurement failure. KFC had switched its UK delivery contract to DHL—a logistics company brilliant at moving things across the globe but with zero experience in cold-chain food distribution. The transition went badly. Deliveries piled up. Distribution centers got overwhelmed. Chicken rotted. Nearly a thousand restaurants sat empty, staffed by employees with nothing to do and customers with nowhere to eat. The financial hit was about £1 million per day. The reputational damage? Incalculable.

One bad sourcing decision, one poorly managed supplier transition, one boardroom moment where somebody probably said, "Yeah, they can handle it," without doing the proper due diligence—and suddenly KFC is the butt of every joke on the internet and real human beings are calling police stations to report a chicken shortage.

Welcome to procurement.

GREAT SCOTT. WHAT IF WE'D HAD AI?

Let's run this back. The chicken story you just read is real, and it happened in February 2018. What would have happened if the same procurement team had the tools that exist now? Not the future—now. The tools you can sit down with on a Tuesday afternoon and use.

KFC's UK procurement team made three decisions that, in retrospect, are easy to second-guess. They chose DHL, built a transition plan, and went live. Let's take each one.

DHL got the contract for a reason. They're brilliant at global logistics. What they had never done was cold-chain food distribution at the scale of nearly nine hundred quick-service restaurants depending on daily deliveries. A procurement analyst in 2017 had every right to know this. The information was public—DHL's case studies, customer references, news coverage, industry analyst reports. The constraint was time. Building a supplier capability assessment against the operational requirement, at the depth needed to surface the cold-chain gap, was the kind of work that took two analysts the better part of a week and got skipped half the time because nobody had the bandwidth.

What does that work look like with an AI tool and a competent analyst at a keyboard? About forty-five minutes. The prompt is roughly: "I'm evaluating [vendor] for [operational scope]. Synthesize their public capability documentation, customer references, recent press, and any disclosed operational incidents into a capability gap assessment against the following requirements." The output isn't a recommendation. It's a structured map of where the supplier has done this before, where they haven't, and what the financial press and customer references suggest about their gap-closure track record. The KFC team would have seen the cold-chain gap on the first read.

Now the transition plan. AI doesn't build the plan. The

people do. But AI runs the kind of adversarial stress-test that good procurement teams have always wanted to do and rarely had time for. The prompt: "Act as DHL operations. You've just been awarded the KFC UK delivery contract. Walk me through what your first major operational failure looks like, when it happens, and how it propagates. Be brutally honest about the things you'd normally not tell the customer." A category manager who runs three rounds of that war-gaming, the morning before the contract is signed, surfaces the exact failure mode that ended up running on every news channel in Britain. Maybe they call DHL with sharper questions, restructure the transition in stages, or keep the old supplier on standby for ninety days. They don't avoid the risk. They price it.

The third decision was going live. AI can't help you make that call. What it can do is monitor the rollout in real time—pulling restaurant-level delivery confirmations, social media chatter about individual locations, manager complaints surfacing on internal channels, news mentions, traffic to KFC's own website—into one signal feed that screams when something is wrong. The KFC crisis didn't blow up on a Sunday afternoon. It built across forty-eight hours during which procurement at corporate didn't know how bad it was. An AI monitoring layer would have known within four hours.

None of this is magic. None of it would have guaranteed a different outcome. KFC had warning signs they could have caught with a clipboard and a phone, if anyone had time to call. What AI changes is that the work is now possible to do—the capability gap analysis, the adversarial war-game, the real-time monitoring. All of it is a Tuesday afternoon for someone who knows what they're doing.

The chicken might still have rotted. The reputation hit might still have happened. But the procurement team would have known why.

Now, if you think that story is bad, let me introduce you to one that's less funny and more expensive. Let's talk about Takata.

Takata was a Japanese auto-parts maker. Specifically, they made airbags—the explosive devices inside your steering wheel and dashboard that deploy in a fraction of a second during a collision and save your life. For years they were one of the dominant suppliers in the automotive world, shipping airbags to Honda, Ford, BMW, Toyota, Nissan, Chrysler, and most of the other major automakers on the planet. A reliable, trusted supplier. And then, slowly and catastrophically, it became clear that their airbags were defective.

Defective in a way that's hard to overstate: under conditions of moisture and temperature fluctuation, the metal inflator would rupture—turning the device designed to save your life into a shrapnel cannon aimed at your face.

The consequences were devastating. As of this writing, Takata's airbags have been linked to at least twenty-seven deaths and hundreds of injuries worldwide. The recall (more than 100 million vehicles globally, across dozens of brands) was the largest automotive safety recall in history. Honda alone paid over $860 million in settlements, fines, and remediation costs. Takata filed for bankruptcy in 2017 with $9 billion in liabilities.

Nine billion dollars.

And here's the part that keeps supply chain professionals up at night: the root cause wasn't just a manufacturing defect that slipped through quality control. It was a procurement and supplier-management failure. Automakers had concentrated their risk in a single supplier. They hadn't conducted adequate audits, demanded sufficient transparency into Takata's manufacturing processes, or built the kind of supplier relationships that produce early warnings of quality degradation. In procurement parlance, they were single-sourced and badly under-managed.

One supplier relationship managed poorly. Over 100 million vehicles recalled. Billions in losses. Twenty-seven people dead.

This is procurement.

THE WARNING SIGNS WERE ALREADY THERE

The Takata story is different from KFC's. It didn't happen in forty-eight hours. It took fifteen years. The first defective inflator was manufactured in 2000. The first death was in 2009. The first major recall wave was in 2013. The full scope of the failure didn't become clear until 2017. Twenty-seven people died and over a hundred million vehicles got recalled because the warning signs accumulated for nearly two decades in fragments that nobody was synthesizing.

That sentence is worth reading twice. The warning signs were there. They were in patent filings, in regulatory disclosures, in NHTSA complaint data, in industry technical journals, in academic studies of propellant chemistry, in Takata's own quarterly financial pressure. They were there. They were just scattered across dozens of data sources, in fragments, none of which constituted a smoking gun on its own.

Synthesis is what AI is for.

Here's what an automaker's risk function looks like with AI capability. A continuous monitoring layer parses every NHTSA complaint, every recall database update, every quarterly financial filing from major suppliers, every academic publication on automotive safety, every industry trade journal. It clusters the data by supplier, by component, by failure mode. It flags anomalies—clusters of complaints from regions with high humidity, financial pressure at a key supplier coinciding with a cost-cutting product change, patent filings that reveal a shift in materials. The clusters become signals. The signals become questions. The questions become audits.

This isn't hypothetical. This is what a competent procurement and risk function builds when the synthesis problem stops being prohibitive.

Could that capability have caught Takata earlier? Honestly, probably yes. The moisture-degradation signal was visible in the data five years before the major recalls. Multiple researchers had published on the chemistry of ammonium nitrate inflators. Takata's financial pressure was visible in their quarterly filings. The concentration of Takata as a single-source supplier across most major OEMs was visible in their public supplier disclosures. Each fragment, on its own, looked like normal industry noise. Together, they were a pattern.

I'm not going to claim AI would have saved twenty-seven lives. That's a claim no honest procurement person should make. What I will claim is that synthesis at the scale required to catch Takata earlier is now within reach of any procurement organization willing to build the capability. The warning signs that took fifteen years to coalesce into action could have coalesced in two.

Same principle. New capability.

I start every conversation about procurement (Rutgers MBA students, Fortune 500 executive rooms, the dinner party guest who just asked me "so what do you do for a living?") with some version of those two stories. Because people picture procurement as the boring back-office function where someone processes purchase orders and argues with vendors about payment terms. If you asked the average MBA which corporate disciplines win or lose markets, procurement would land somewhere between facilities and fleet management on their list. They picture a procurement professional sitting in a beige office, surrounded by stacks of contracts, eating a sad desk lunch while updating a spreadsheet. They don't picture a strategic discipline that wins wars, builds companies, and when done poorly, kills people and bankrupts global enterprises.

My job in this chapter, and really in this entire book, is to disabuse you of that notion.

Because here's what every operator who has built durable competitive advantage already knows: procurement is strategy. Sourcing is power. Supply chain is destiny. The companies that understand this weaponize disruption —they don't just survive it. The companies that don't understand it get blindsided, and they end up starring in the kind of case study business school professors use to humble overconfident students.

They're wrong, and they've been wrong for a long time.

Procurement decisions determine which companies have product when their competitors don't. They determine which automakers spend a decade and ten billion dollars unwinding a single bad supplier choice. They determine, in genuinely consequential ways, who lives and who dies. And over the next five years, they will determine which procurement organizations build a structural advantage with artificial intelligence—and which spend the rest of the decade trying to catch up.

Two arguments run through this book in parallel. The first: the principles that make procurement great have not changed in fifty years and will not change in the next fifty. The second: the capability now available to any procurement professional with an AI tool, an internet connection, and the discipline to use them well is unlike anything the field has seen before.

Both are true.

So let's start at the beginning. What procurement actually is, why it matters, and why the principles that govern it have changed far less than the technology around it.

WHAT IS PROCUREMENT, REALLY?

At its most fundamental level, procurement is the process by which an organization acquires the goods, services, and resources it needs to operate and compete. That definition sounds simple—almost mundane. But embedded in it is an enormous amount of complexity, strategy, risk, and opportunity.

Think about every single thing your organization uses in the course of doing business. The computers your people type on. The software that runs your operations. The raw materials that go into your products. The consultants you hire to help you solve problems. The electricity that powers your buildings. The shipping services that move your goods. The marketing agencies that promote your brand. The legal firms that protect your interests. Every one of those things (every service, every product, every resource) had to be identified, sourced, evaluated, negotiated, contracted, received, and managed by someone. That someone, and the process they follow, is procurement.

Quick definitional detour, because these words get used loosely and the differences matter:

Purchasing is the transactional act of buying—placing orders, processing invoices, making payments. It asks: how do we execute the buy?

Sourcing is the strategic process of identifying, evaluating, and selecting suppliers. It asks: who should we buy from, and why?

Procurement is the umbrella covering both, plus everything that wraps around them—need identification, contracting, supplier performance management, value realization. Procurement asks: how do we manage the entire lifecycle of acquiring what we need?

Supply chain management is broader still. It encompasses procurement and adds the inbound and outbound logistics, operations, inventory, and distribution networks that move goods from origin to end consumer. If procurement is the art of deciding where to buy and managing those relationships, supply chain management is the science of moving everything around once those decisions are made.

For this book, we're using "procurement" in its broadest, most strategic sense—sourcing strategy, supplier management, contract negotiation, risk management, value realization. The way great procurement professionals actually think about it: not a series of transactions, but a continuous competitive-advantage engine.

SUPPLY CHAIN AS COMPETITIVE MOAT

Let's fast-forward fifty years and change the battlefield from Normandy to Bentonville, Arkansas.

In the 1980s and 1990s, Walmart did something its retail competitors couldn't comprehend. It became the most dominant retailer in the history of American commerce (not primarily because it had better products, not because it had better store locations, not because it outmarketed its competitors) but because it built a procurement and supply chain operation that was structurally superior to anything its competitors could match.

Walmart pioneered supplier partnership programs that were, at the time, revolutionary. They invested in information technology to share real-time sales data with suppliers, a practice called Vendor Managed Inventory, so that suppliers could replenish stock proactively, before shelves went empty. They built a private trucking fleet and a network of distribution centers that moved goods from supplier to shelf faster and more cheaply than anyone in the industry. The cost discipline that ran underneath all of it was, frankly, religious.

The result? Walmart could sell at prices competitors couldn't match—not because they had thinner margins, but because their procurement and supply chain costs were so much lower that they could undercut competitors and still make money. Their entire business model, "Every Day Low Prices", wasn't a marketing strategy. It was a procurement strategy—the low prices were the output, the supply chain was the engine.

Apple is the next case, and it's a more surprising one. The brand is built on design and user experience—what most people picture when they hear "Apple" is the iPhone in a black turtleneck on a stage. But Tim Cook, before he ran Apple, ran its supply chain. And the operation he built is one of the most sophisticated in the world: locking up production capacity at suppliers years in advance, managing hundreds of component vendors across dozens of countries, coordinating new-product launches at global scale with what observers have charitably called military precision. These are procurement and supply chain capabilities that have given Apple structural advantages competitors have spent years and billions trying to replicate.

Amazon is the third case, and the most direct. It's commonly called a tech company that sells things. It might be more accurately called a logistics company that happens to have a great website. When you click "buy now" and your package arrives the next morning, sometimes the same afternoon, that isn't magic. It's the output of a procurement and logistics operation of jaw-dropping complexity, working flawlessly in the background.

The pattern is unmistakable. Some of the most successful companies in commerce, companies that look on the surface like they're about retail or technology or consumer electronics, are at their core procurement and supply chain machines. The supply chain is the moat. The procurement capability is the engine. The companies that invest in that capability, that treat it as a strategic priority rather than an administrative function, build competitive advantages that last for decades.

THE HIDDEN COST OF GETTING PROCUREMENT WRONG

We've looked at the spectacular failures—KFC's chickens, Takata's airbags—and the spectacular successes—Walmart, Apple, Amazon. But most procurement failure isn't spectacular. It's quiet, persistent, and expensive precisely because nobody notices it happening.

Some numbers that should make every executive uncomfortable. Research consistently shows that organizations with mature, strategic procurement functions outperform their peers on profitability by three to six percentage points. On a billion-dollar revenue base, that's $30 to $60 million in annual profit difference—not from one brilliant negotiation, but from the cumulative impact of thousands of better-informed sourcing decisions, tighter supplier management, more rigorous contract administration, and more effective risk mitigation, compounding year after year.

The inverse is grimmer. Organizations that treat procurement as an administrative function (staffing it with whoever is available, measuring it on price reduction rather than value creation, underinvesting in technology and capability) aren't just missing the upside. They're hemorrhaging value through mechanisms that are individually modest and collectively staggering.

They're paying more than they should for goods and services because nobody has the time or the data to benchmark their pricing against the market. Contract value is leaking through auto-renewals nobody tracked, escalation clauses nobody monitored, and service levels nobody measured. They're carrying supplier risk nobody has assessed because the risk-management process is an annual questionnaire rather than a continuous monitoring capability. They're making sourcing decisions based on whoever the business unit happens to know rather than a disciplined evaluation of the full market. And they're missing innovation opportunities because their supplier relationships are too transactional to surface them.

None of this makes the evening news. None of it triggers a hashtag or a government statement. But it adds up to a competitive disadvantage that compounds—and that the organizations on the other side of the equation are exploiting relentlessly.

This is why I get frustrated, genuinely frustrated, when I encounter organizations that still treat procurement as a back-office function. Not because it offends my professional pride, although it does. But because I can see, from thirty years of experience, exactly how much value they're leaving on the table. The gap between what procurement could deliver and what most organizations allow it to deliver is, in my estimation, the single largest untapped competitive advantage in most enterprises.

The companies that have figured this out (the Walmarts, the Apples, the Amazons, the Toyotas) are not doing something mystical. They're doing procurement well. Strategically, rigorously, with investment and executive attention and the best talent they can find. The principles are not secret. The frameworks are not proprietary. The difference is commitment—the organizational commitment to treat procurement as the strategic function it genuinely is.

That commitment is what this book is about.

THE FUNDAMENTALS THAT NEVER CHANGE

Here's something that might surprise you, given that I'm writing a book about how AI is transforming procurement: the principles that govern great procurement haven't changed. Not in any meaningful way. The tools have evolved enormously. AI is creating capabilities that would have seemed like science fiction twenty years ago. But the principles? Five of them. Named the same way I've been hearing them named for thirty years.

Value, not price. The lowest sticker doesn't mean the lowest total cost. A supplier who's twenty percent more expensive but lasts twice as long and never makes you fight about service levels is cheaper than the one who beat them on the opening number.

Risk is always present. The automakers who got

crushed by Takata weren't stupid. They were competent procurement organizations that had not adequately mapped, quantified, and managed their supplier risk. Risk doesn't go away because nobody noticed it yet.

Relationships are assets. The customer who has built genuine partnership with their critical suppliers gets calls at eleven at night during a shortage. The customer who has spent three years pounding the table on price gets put on the back of the allocation queue.

Information is power. Always was—AI hasn't changed that. It's changed what's possible to gather and synthesize.

Process discipline creates results. Great procurement isn't about having one brilliant person. It's about building processes that consistently deliver good outcomes whether the brilliant person is in the room or not.

These five principles are the ground floor. Internalize them, and AI will multiply your impact. Skip them, and AI will be ornamentation on a function that still doesn't work.

Let's keep going.

CHAPTER 1
FROM BEAT-DOWN TO BUILD-UP

THE EVOLUTION OF SUPPLIER RELATIONSHIPS

Let me tell you about my boss.

It was the early 1990s. I was young, I was hungry, and I was learning the procurement trade from a man who had elevated the art of vendor intimidation to something approaching a performance. His name is Big Ron, It was on his business card, it was his email address, and if you've spent more than six months in procurement you've either worked for one, worked alongside one, or, and I say this with genuine affection, been one yourself at some point.

Big Ron was not a subtle man. His philosophy of supplier management could be summarized in approximately seven words: find out what they need and use it. If a vendor needed the contract to make their quarterly numbers, Big Ron knew it. If a vendor's regional rep was under pressure from their VP to close the deal before month-end, Big Ron had already done the math on exactly how much additional concession that pressure was worth. If a supplier was single-sourced and therefore had no credible competition, Big Ron would manufacture competition (sometimes real, sometimes theoretical, always strategically deployed) just to keep them honest and uncomfortable.

His negotiation sessions were legendary. He would walk into a conference room like a man who had already won and was simply waiting for the other party to figure it out. He would ask for things he didn't need just so he had something to give back. He would go silent at precisely the right moment (that excruciating, loaded silence that amateur negotiators always break too soon) and wait for the supplier's side of the table to start making concessions out of sheer discomfort. He would compliment the vendor's product extravagantly one minute and then question whether they could really deliver at scale the next, leaving the supplier representative in a state of perpetual uncertainty about where they stood.

Was he effective? In the short term, absolutely. Big Ron saved money, hit his numbers, and got the concessions.

Was he building anything? No. Not really. And that distinction, between extracting value and building value, is the central thesis of this entire chapter.

Before we go further, a word for the senior leaders in the room. The shift this chapter describes, from adversarial vendor management to genuine supplier partnership, is not a procurement-cultural fashion. It's a structural change in how supply-side competitive advantage gets built. The organizations that figured this out twenty years ago have, on average, healthier supply chains, more resilient revenue, and access to supplier innovation their competitors can't buy. The ones that didn't are still squeezing margin out of relationships that are quietly squeezing back. The mechanics in this chapter (segmentation, performance management, joint planning, supplier development) are tactical. The argument underneath them is strategic. And in an environment where geopolitical, climate, and AI-driven disruption is the new baseline, that argument is more decisive than it has been at any point in my career.

THE PROCUREMENT CLIMATE OF THE 1990S: EVERYBODY WAS BIG RON

To be fair to Big Ron, and to be fair to the era, this approach to supplier management wasn't an aberration. It was the dominant paradigm. The prevailing wisdom in corporate procurement through most of the 1980s and into the 1990s was fundamentally adversarial. Suppliers were not partners, collaborators, or strategic assets. They were vendors—a word that, in certain procurement circles, was practically a pejorative—sources of supply to be managed, which in practice meant controlled, pressured, and periodically reminded of their replaceability.

The procurement function itself was largely measured on one dimension: cost savings. How much did you spend last year? How much did you spend this year? The difference between those two numbers was your performance. The tools for generating that savings number were limited —competitive bidding to drive prices down, consolidation of spend to create leverage, hard-nosed negotiation to extract concessions. Supplier relationships, in this framework, were valuable primarily as a source of future leverage. You maintained them just well enough to ensure continued competition, but not so well that suppliers got comfortable.

This wasn't malicious. It was rational, given the metrics and incentives of the time. If the only thing your organization measures is cost reduction, and the primary mechanism for cost reduction is supplier pressure, then maximizing supplier pressure is the logical response. Big Ron wasn't a bad person. He was a rational actor in a system that rewarded exactly what he was doing.

The problem, and it took the procurement profession a surprisingly long time to fully reckon with this, is that the adversarial model has a ceiling. You can only squeeze so much out of a supplier relationship before the relationship breaks or the supplier finds ways to recover their margins that you can't see. And once you've spent years treating suppliers as adversaries, you've systematically destroyed the very thing that creates the most long-term value in a supply relationship: trust.

I watched this play out more times than I can count. A supplier who's been beaten up on price for three consecutive contract cycles starts finding ways to make the numbers work that aren't necessarily in your best interest. Specifications get interpreted as narrowly as possible. Service levels drift to the contractual minimum rather than the practical maximum. When you need something outside the standard scope (an urgent delivery, a custom solution, a creative accommodation) the supplier's flexibility evaporates.

Why would they go above and beyond for a customer who has spent years treating them like an opponent? They wouldn't. And they didn't.

And then, gradually, the world changed.

THE TOYOTA SHOCK: A DIFFERENT WAY WAS POSSIBLE

If there is a single organization most responsible for catalyzing the shift in how the Western business world thought about supplier relationships, it's Toyota. The story of how Toyota's supplier philosophy eventually penetrated American and European procurement thinking is one of the most important intellectual evolutions in the history of the profession.

Toyota's approach to suppliers, formalized as part of the broader Toyota Production System that the West would later popularize as "lean manufacturing", was fundamentally different from the adversarial model in almost every respect. Where Western companies maintained large competitive supply bases and played suppliers against each other, Toyota deliberately concentrated its supply base, working with a smaller number of carefully selected suppliers with whom it built deep, long-term relationships. Where Western procurement was focused on minimizing purchase price, Toyota focused on minimizing total system cost—working with suppliers to eliminate waste, improve quality, and reduce cost throughout the entire supply chain rather than simply transferring cost pressure onto the supplier side of the ledger.

Most distinctively, Toyota invested in its suppliers. When a Toyota supplier had a quality problem, Toyota would send its engineers to help solve it—not as a punitive measure, but as a collaborative effort to fix the root cause together. Toyota shared production forecasts with suppliers far in advance so they could plan their capacity efficiently. Toyota helped suppliers develop their capabilities, implement lean practices, and build the organizational competencies that would make them better partners. In return, Toyota expected, and received, extraordinary levels of performance, flexibility, and commitment.

The results were undeniable. Toyota suppliers consistently outperformed the suppliers of Toyota's American and European competitors on quality, delivery, cost, and innovation. Toyota's supply chain was more resilient, more responsive, and more capable of continuous improvement than anything the traditional adversarial model could produce.

And when a major disruption hit, the 1997 Aisin Seiki fire that destroyed a key brake parts factory and threatened to shut down all of Toyota's Japanese assembly plants, the depth of Toyota's supplier relationships created a response capacity that was, frankly, hard to believe. Dozens of other Toyota suppliers, some of whom didn't normally make brake parts, voluntarily retooled their operations to help cover the production gap. Not because they were contractually obligated. Because they were partners.

That's not something you get from a vendor you've

been beating up for years. That's the dividend of relationship investment, paid precisely when you need it most.

THE PARADIGM SHIFT: FROM VENDOR MANAGEMENT TO SUPPLIER RELATIONSHIP MANAGEMENT

The formal articulation of Supplier Relationship Management, SRM, as a distinct professional discipline emerged in the late 1990s and accelerated through the 2000s, driven in part by the growing recognition of Toyota's model, in part by the increasing complexity and interdependence of global supply chains, and in part by a series of supply chain disruptions that exposed how costly the adversarial model's brittleness could be.

SRM, at its core, represents a fundamentally different answer to the question: what is the purpose of managing supplier relationships? The adversarial model answered that question with "to extract maximum value from the supplier in each transaction." SRM answers it with "to develop and maintain supplier relationships that create sustainable, growing value for both parties over time."

That shift (from extraction to creation, from transaction to relationship, from short-term to long-term) changes almost everything about how you interact with suppliers. Let's walk through what modern SRM actually looks like in practice.

SUPPLIER SEGMENTATION: NOT ALL SUPPLIERS ARE CREATED EQUAL

The first and most foundational concept in SRM is supplier segmentation—the recognition that different suppliers play different roles in your organization's value creation, and therefore deserve different levels of investment and different styles of relationship management.

This might seem obvious, but its implications are more profound than they first appear. Most organizations, if you look at their actual procurement behavior, treat all suppliers more or less the same—the same arm's-length transactional approach, the same focus on price, the same periodic competitive rebid whether it makes strategic sense or not. This is procurement as administrative function, not as strategic discipline.

The segmentation model starts with a simple but powerful observation: in most organizations, a relatively small number of suppliers account for a disproportionately large share of either spending or operational criticality. Some suppliers provide inputs so essential to your operations (components with no readily available substitute, services that would be catastrophically disruptive to lose) that the relationship itself is a strategic risk factor that requires executive attention. Other suppliers provide commoditized goods or services where the relationship is transactional by nature and the primary management lever is competitive pricing. Most suppliers fall somewhere in between.

The classic segmentation framework (known in procurement education as the Kraljic Matrix, after Peter

Kraljic's foundational 1983 Harvard Business Review article) plots suppliers on a two-by-two: supply risk (or criticality) on one axis, spend value (or strategic importance) on the other. The four quadrants drive different relationship models. High spend, high criticality suppliers become strategic partners—dedicated relationship management, joint business planning, executive engagement, real mutual investment. High spend, lower criticality suppliers are typically managed through preferred-supplier programs focused on performance and commercial optimization. Lower spend, high criticality suppliers (the bottleneck quadrant) require careful risk management even when the dollar amounts are modest. And lower spend, lower criticality suppliers are managed transactionally—efficient, arms-length, appropriately automated.

The segmentation framework does two critical things. First, it ensures that your most important supplier relationships receive the investment they deserve rather than being managed with the same one-size-fits-all approach as your office supply vendor. Second, it provides a rational, defensible basis for differentiating your relationship approach—so that when you're being a genuine strategic partner to a key supplier, that supplier understands the elevated relationship reflects their strategic importance and is not simply you trying to soften them up before the next negotiation.

WHAT STRATEGIC PARTNERSHIP ACTUALLY LOOKS LIKE

The phrase "strategic partnership" gets thrown around so promiscuously in business that it has lost much of its meaning. Everyone is a "strategic partner" in the brochure. In practice, real strategic supplier partnerships are relatively rare—and they look quite different from conventional supplier management.

Let me describe what one looks like, drawn from my own experience.

Early in my career at a major Fortune 500, we had a technology supplier that mattered. Their platform ran a business-critical workflow at enterprise scale—high seven-figure annual spend, embedded so deeply in our operations that replacing it would have been an eighteen-month, multimillion-dollar project. By every measure on the segmentation matrix, they were a strategic partner.

For years, we hadn't managed the relationship like one. Contract negotiations were adversarial. Quarterly business reviews were essentially complaint sessions where we catalogued their service failures and they explained why the issues were our fault or outside their control. There was minimal transparency on either side—we didn't share our roadmap with them, they didn't share theirs with us, and the relationship had all the warmth and collaborative energy of a hostage negotiation.

Then it changed—not because either party had an epiphany, but because we got a new procurement leader with a different philosophy, and because the supplier's account team had decided they were tired of the adversarial dynamic too.

The change was gradual but transformative. We started sharing strategic plans with the supplier—not everything, but enough that they could align their development priorities to what we were going to need. They started giving us early access to their product roadmap—not as a sales pitch, but as a real conversation about whether their direction matched ours and where there were gaps. We established a joint steering committee with executive participation on both sides that met quarterly to discuss strategic alignment rather than operational issues. We moved from a transactional pricing model to a multi-year agreement that gave them revenue predictability in exchange for pricing stability and preferential service levels.

And here's what happened. The quality of the solution improved dramatically. Because the supplier actually understood what we were trying to accomplish, they started developing features and capabilities we needed rather than features that were easy to build or that the sales team thought they could sell. When we had operational problems, they were invested in solving them because we were invested in their success. When we had an urgent need outside our contract scope, they accommodated it—because the relationship had built a reservoir of goodwill that made flexibility possible.

The financial outcome was the part most people don't expect. Our total cost of ownership went down—not because we had beaten the supplier into submission on price, but because the partnership had eliminated so much waste from the relationship: the waste of adversarial negotiations, the waste of performance problems that lingered because nobody was invested in the root cause, the waste of missed opportunities because we hadn't shared our needs clearly enough.

That's what strategic partnership looks like when it's real. Not a plaque on the wall and a logo on a slide deck. Genuine alignment of interests, transparent communication, mutual investment, and shared accountability for outcomes.

THE TRUST EQUATION: WHAT IT TAKES TO BUILD, WHAT IT COSTS TO BREAK

Everything we're about to walk through (the architecture of modern SRM, the performance frameworks, the commercial structures, the joint planning and supplier development practices) sits on top of a single foundation: trust. Without it, the architecture doesn't hold.

The procurement professionals who are most effective at building real supplier partnerships understand something the Big Ron school never quite grasped: trust is not naivety. Being a trustworthy and trusting business partner does not mean being soft. It does not mean abandoning commercial discipline. It does not mean accepting poor performance or overpaying for goods and services. It means being honest about your needs and constraints, keeping your commitments, dealing with problems directly and respectfully rather than defensively and punitively, and treating the people on the other side of the table as professionals deserving of respect regardless of which side of the contract they sit on.

The practical implications are more far-reaching than they might appear. It means that when you tell a supplier you're running a competitive process and will award on merit, you actually do that—you don't conduct a fake competition and award to the incumbent for political reasons, leaving the other bidders wondering why they wasted their time. It means that when you commit to a payment schedule, you meet it—because the suppliers experiencing your 90-day payment terms while you're calling them a strategic partner are not confused about what you actually value. It means that when a supplier brings you bad news early (a quality problem, a capacity constraint, a pricing pressure they can't absorb) you respond with problem-solving rather than punishment, because the alternative is a supply base that buries bad news until it becomes a crisis.

Trust in supplier relationships is also, critically, an institutional and not just a personal quality. The supplier who trusts you as a person but has been burned by your organization's procurement process (ambushed in a negotiation, re-bid without warning, subjected to payment delays) does not have a trusting relationship with your organization. Building organizational trust requires not just the right individual behaviors but the right systems and processes: clear, consistent, fair sourcing practices; reliable commercial processes; accessible escalation paths when problems arise; and leadership that models partnership values rather than espousing them.

THE ARCHITECTURE OF MODERN SRM: KEY PRACTICES AND FRAMEWORKS

With trust as the foundation, the practices that govern how SRM gets executed at every level make sense as something more than process for its own sake. Let me walk through the most important ones.

Supplier Performance Management

Every meaningful supplier relationship should be governed by clearly defined performance metrics, KPIs, agreed at the outset of the relationship and tracked consistently over its life. The specific metrics vary by category. A logistics provider might be measured on on-time delivery, damage rates, and invoice accuracy. A software vendor might be measured on uptime, response times, and resolution rates. A professional services firm might be measured on project delivery against milestones, quality of deliverables, and stakeholder satisfaction.

The purpose of performance metrics is not to generate ammunition for negotiations—though performance data does legitimately inform commercial discussions. The purpose is shared visibility into how the relationship is actually performing, early identification of problems while they're still solvable, and a foundation for continuous improvement rather than episodic crisis management.

Critically, performance management is a two-way street in a real partnership model. Just as the supplier is accountable for performance against agreed metrics, the buying organization should be willing to evaluate its own performance as a customer—whether it provides forecasts on time, whether its payment processes are reliable, whether its requirements are clear and stable, whether its people treat the supplier's team with professional respect. The organizations willing to hold themselves accountable as customers, measuring themselves on what some of the best buyers call "customer of choice" metrics, are the ones that attract the best performance from their suppliers.

Quarterly Business Reviews

The Quarterly Business Review, or QBR, is the standard vehicle for structured relationship management in most organizations—and it is also, in many organizations, a ritual of remarkable ineffectiveness. Badly run QBRs are essentially one-party complaint sessions: the buying organization presents a slide deck of performance failures, the supplier defends themselves, nothing is resolved, and everyone leaves having spent two hours accomplishing nothing except mutual reinforcement of the adversarial dynamic.

Good QBRs are fundamentally different in structure and purpose. They are bilateral—both parties bring their perspective on the relationship, including the buying organization's assessment of how it has been performing as a customer. They are forward-looking as well as backward-looking—spending at least as much time on where the relationship is going as on where it's been. They are attended by people with decision-making authority—not just relationship managers, but executives on both sides who can actually commit to changes and investments. And they are structured around shared goals rather than competing grievances.

I've sat in hundreds of QBRs over the course of my career. The ones that actually moved relationships forward shared a common characteristic: they felt like conversations between colleagues trying to solve shared problems, not interrogations of an underperforming subordinate. The format matters less than the mindset—but having an agenda structured around shared value creation rather than supplier accountability goes a long way toward establishing the right mindset before anyone walks in the room.

Joint Business Planning

For the most strategic supplier relationships, QBRs are supplemented, sometimes replaced, by something more ambitious: joint business planning. JBP is the practice of sitting down with a strategic supplier annually or semi-annually to align your respective business plans and identify the areas of mutual opportunity and interdependence.

A well-executed joint business plan answers questions like: what are each party's key business objectives for the coming year, and where do those objectives intersect? What investments is the supplier planning in capability or capacity that will affect the buying organization? What requirements is the buying organization planning that will affect the supplier's priorities? Where are there innovation opportunities neither party could realize alone? What are the shared risks both parties need to plan for?

The output is not a document—it's a shared understanding and commitment that replaces the unspoken guessing games governing most supplier relationships with explicit alignment. And explicit alignment, it turns out, is enormously valuable. The suppliers who know what you need before you ask for it, who have planned their capacity around your growth trajectory, who have invested their R&D resources in the problems you're trying to solve—those suppliers deliver qualitatively better outcomes than the ones who are perpetually surprised by your requirements.

Supplier Development

Perhaps the most advanced expression of the partnership model is supplier development—the practice of actively investing in a supplier's capabilities, not just managing their existing performance.

This might mean providing technical assistance to help a supplier solve a persistent quality problem. It might mean connecting a supplier with other customers who have expertise in process areas where the supplier is weak. It might mean co-investing in a supplier's capacity expansion because your projected volume growth makes that expansion mutually beneficial. It might mean helping a supplier navigate a certification requirement (an ISO quality standard, a regulatory compliance requirement, a cybersecurity framework) that your organization needs them to meet.

Supplier development requires a level of trust and openness that is the antithesis of the adversarial model. It requires being willing to show suppliers your own vulnerabilities (where you're dependent on them, where your own processes create problems for them) because you can't fix a supplier's capability gap if you're pretending you don't have one. It requires a long-term orientation, because the payoff from supplier development investments typically comes over years, not quarters. And it requires real partnership rather than performative partnership, because suppliers will not invest in the relationship if they believe they'll be re-bid the moment a cheaper alternative emerges.

The organizations that do supplier development well are building competitive advantages that are genuinely hard to replicate. The supplier capability you've invested years in developing is not available to your competitors. The depth of understanding and alignment that comes from sustained co-investment is not something a competitor can acquire by switching to your supplier. It's a structural advantage embedded in relationships that took years to build—and that's exactly why it's worth building.

THE COMMERCIAL ARCHITECTURE: GETTING INCENTIVES RIGHT

One of the most important insights in modern SRM is that the structure of commercial arrangements (how you pay suppliers, how you share risk and reward, how you define the terms of the relationship) has enormous influence on supplier behavior. Get the incentives right, and suppliers are naturally aligned with your interests. Get them wrong, and you'll spend years trying to manage behavior your own contract is actively incentivizing.

The classic example of misaligned incentives is the time-and-materials contract for professional services. In a T&M arrangement, the supplier is paid for hours worked—so what incentive do they have to work efficiently? Exactly the wrong one. Under a pure T&M model, the supplier is economically rewarded for taking longer, not for delivering faster. Smart buyers recognize this and structure contracts that align supplier incentives with buyer outcomes: fixed-price arrangements for well-defined scopes, outcome-based pricing for value creation, gain-share mechanisms for cost reduction or revenue generation initiatives.

The gain-share model deserves particular attention because it's, in commercial terms, the clearest expression of the partnership philosophy. In a gain-share arrangement, the supplier is compensated not just for delivering a service but for delivering measurable value—and the economic upside of that value creation is shared between the buying organization and the supplier. This alignment of economic interests transforms the supplier from a service provider managing to a contract into a partner invested in your success.

I've implemented gain-share arrangements in several categories over my career (technology transformation projects, logistics optimization initiatives, procurement cost reduction programs) and the difference in supplier behavior compared to traditional cost-plus or fixed-price arrangements is striking. When the supplier has real economic skin in the game, when their team's bonus depends on whether your project succeeds, the conversation changes from "are we meeting the contract terms?" to "are we creating the outcome we both committed to?" That's a fundamentally different and more productive conversation.

THE SUPPLIER RELATIONSHIP IN MODERN PRACTICE: NEW TOOLS, SAME HUMAN TRUTH

Before we go further, I want to address something that comes up every time I teach this material—whether it's in a Rutgers classroom, a corporate training session, or a conference room full of executives who've just had their third cup of coffee and are starting to get skeptical. Someone always raises their hand, or their eyebrow, and asks the version of this question: "David, this all sounds great. But isn't this just procurement people being soft? Aren't we giving up negotiating leverage by getting too friendly with our suppliers?"

It's an excellent question, and the answer is no—but the reasoning matters.

The partnership model does not mean you stop negotiating. It does not mean you accept whatever price a supplier quotes because you've decided they're a strategic partner and it would be awkward to push back. It does not mean you abandon competitive sourcing processes, ignore market benchmarks, or give incumbent suppliers a free pass at renewal because the relationship is comfortable. Any procurement professional who interprets "partnership" as "softness" has fundamentally misunderstood the model.

What the partnership model actually means is that you negotiate differently, not less. You negotiate with better information—because a supplier who trusts you shares cost structure, margin pressures, and operational constraints in ways an adversarial supplier never would, and that information makes you a much more effective negotiator. You negotiate with a longer time horizon—optimizing for total value over the life of the relationship rather than squeezing every last dollar out of each individual transaction, which frequently costs more in the long run than it saves in the short term. And you negotiate with your eyes open to the full value equation—understanding that price is one variable in a more complex optimization problem that includes quality, reliability, innovation, risk, and relationship capital.

The best procurement negotiators I've ever encountered, the ones who consistently got the best outcomes for their organizations over time, were not the hardest or most aggressive. They were the best prepared. They knew more about the market than the supplier expected them to know. They understood the supplier's business better than the supplier thought any customer would bother to understand it. They came to the table with creative options rather than ultimatums. And they built relationships deep enough that when they said something was a hard requirement, the supplier believed them—because the relationship had a track record of honesty that gave those words real weight.

That's what partnership enables. Not softness. Sophistication.

WHEN SUPPLIER RELATIONSHIPS GO WRONG: THE WARNING SIGNS

No chapter on supplier relationship management would be complete, or honest, without talking about what happens when supplier relationships go sideways. Because they do. Even the most carefully managed partnerships hit rough patches. Suppliers miss deliverables. Quality degrades. Key account team members you relied on get promoted or leave. Commercial pressures change the supplier's priorities in ways that diverge from yours. And sometimes, despite everyone's best intentions, a relationship that started with partnership energy devolves into something transactional, resentful, or outright adversarial.

The warning signs are usually visible well before the relationship reaches crisis point—if you're paying attention. Four are worth watching for in particular.

Declining transparency is the first and probably the most diagnostic. When a supplier who used to proactively share information (forecast challenges, quality concerns, staffing changes on your account) starts going quiet, something has shifted. Silence in a supplier relationship almost never means everything is fine. It usually means the supplier has decided that being honest with you about a problem is more dangerous than trying to manage it without telling you. That's a relationship that has lost trust, and it needs intervention before the hidden problem becomes an unmanageable crisis.

Performance drift is the second, and the most common reason supplier relationships fail in slow motion. Most performance degradation doesn't happen overnight. It happens gradually (metrics slipping from excellent to good, from good to acceptable, from acceptable to below standard) while the supplier finds reasons to explain each slip that seem plausible in isolation. The antidote is rigorous performance tracking that makes the trend visible rather than allowing each individual data point to be rationalized in isolation.

Watch also for escalating commercial friction. Every contract renegotiation involves some level of commercial tension—that's normal and healthy. But when a supplier relationship starts being characterized by constant disputes over scope, persistent arguments about invoices, and a pattern of commercial positions that feel more like tests of leverage than business discussions, something fundamental has broken down.

The last one is the most subtle and the easiest to miss: the loss of problem-solving orientation. Healthy supplier relationships, even ones under stress, share an orientation toward solving problems together. When that shifts, when supplier conversations start feeling like two parties arguing about whose fault something is rather than two parties trying to fix it, the relationship has moved from partnership to adversarialism, and it requires deliberate intervention to reset.

When you see these signs, the worst thing you can do is ignore them and hope they resolve themselves. They don't. The professional move at that point is to surface the issues directly (not punitively, but honestly) and have a real conversation about what's changed, what both parties need, and whether the relationship can be reset to a productive footing.

Sometimes it can. Sometimes the honest answer is that the relationship has run its course and a transition is in order—and recognizing that clearly and managing it professionally is also part of supplier relationship management excellence.

Because here's what the adversarial model gets right, even when it gets almost everything else wrong: you should never be afraid to change suppliers when the business case is clear. Loyalty to a supplier relationship should never become an excuse for sustained underperformance or persistently uncompetitive commercial terms. The partnership model is about building mutual value—and when a supplier is no longer contributing to that value, the professional response is not sentimentality. It's a well-managed transition to a better alternative.

THE SUPPLIER'S PERSPECTIVE: WHAT BEING A GREAT CUSTOMER ACTUALLY MEANS

Here's a perspective shift that changes how procurement professionals think about supplier relationships, and one I don't think gets nearly enough airtime in the procurement literature: what does it actually feel like to be your supplier?

I've spent the vast majority of my career on the buy side. But I've also had enough conversations with suppliers (in negotiations, in QBRs, in the kind of honest post-engagement conversations that happen when everyone is being real) to have developed a pretty clear picture of what it's like to work with different kinds of customers. And the variation is striking.

The best customers (the ones suppliers talk about with real enthusiasm, the ones that get allocated the best talent and the most creative problem-solving energy) share certain characteristics. They're clear about what they need. They don't change their requirements constantly and then act surprised when the supplier can't accommodate the changes. They pay on time. This sounds basic, but you would be stunned by how many large corporations have payment processes so dysfunctional that their suppliers effectively end up providing free financing, sometimes for ninety days or more, while the customer's accounts payable team fights over invoice approvals. They treat the supplier's people with respect. They provide feedback—not just complaints about failures, but real input on what's working and what isn't, delivered in a spirit of helping the relationship improve. And they honor their commitments: when they say they're going to run a competitive process and award on merit, they do it. When they commit to a volume forecast, they do their best to honor it. When they agree to a commercial term in a negotiation, they don't come back two weeks later trying to reopen it.

The suppliers who work with those customers do so for one reason: revenue. And they price the relationship accordingly—charging more for the uncertainty and friction, staffing the account with whoever is available rather than whoever is best, and taking the first opportunity to walk away when a better option presents itself.

Here's why this matters for procurement professionals: you're significantly more powerful than you think in shaping the customer experience your organization delivers to its suppliers. The way you structure RFP processes, run negotiations, manage QBRs, handle payment disputes, communicate changes in requirements —all of these behaviors aggregate into your organization's reputation in the supply market. And that reputation determines the quality of suppliers willing to compete for your business, the quality of talent those suppliers assign to your account, and the quality of discretionary effort those suppliers bring to the relationship.

Being a great customer is not a soft aspiration. It's a competitive strategy.

THE MEASUREMENT CHALLENGE: HOW TO TRACK RELATIONSHIP VALUE

One of the persistent challenges of the partnership model, and one of its vulnerabilities to organizational skepticism, is measurement. The adversarial model had a simple, compelling metric: cost savings. How much did we reduce price? The number was clear, attributable, and easy to communicate to leadership. It might have been misleading (saving ten percent on price while degrading quality, increasing risk, and destroying the supplier's incentive to innovate isn't actually a good outcome) but it was measurable. And in organizations that run on metrics, measurable beats valuable when the two are in conflict.

The partnership model creates value that's harder to measure but no less real. Supplier innovation that would never have been proposed in an adversarial relationship. Risk mitigation through early warning that prevented a disruption. Service flexibility that avoided a production delay. Priority allocation during a shortage that kept your lines running while competitors scrambled. The collaborative problem-solving that identified a root cause and fixed it permanently rather than just managing symptoms.

How do you put a number on the disruption that didn't happen? On the innovation that the supplier brought you proactively? On the priority treatment during a crisis that was earned through years of partnership investment?

The honest answer is that some of these value dimensions resist precise quantification. But that doesn't mean they can't be measured—it means they require a more sophisticated measurement approach than a simple savings calculation.

The organizations that do this well typically use a balanced scorecard approach to supplier relationship value that includes multiple dimensions. Cost performance, yes—but measured as total cost of ownership rather than unit price. Quality performance against agreed metrics. Delivery reliability. Innovation contribution—the number and value of improvement ideas the supplier has brought to the table. Risk management effectiveness—the quality of the supplier's business continuity planning, their transparency about potential issues, their responsiveness when problems arise. And a qualitative assessment of relationship health, conducted bilaterally, with both the buying organization and the supplier evaluating the relationship on dimensions like communication quality, trust, and mutual respect.

This balanced approach doesn't produce a single number that you can put on a slide. It produces a multidimensional picture of relationship value that's more complex but more accurate than a savings metric alone. And it provides the ammunition procurement leaders need when the CFO asks "why aren't we just beating our suppliers up on price?"—because the balanced scorecard shows what partnership delivers that adversarial management doesn't.

Be practical about this. Measurement approaches that are theoretically sound but operationally burdensome don't get sustained. The system needs to be lightweight enough that procurement professionals actually use it and comprehensive enough to capture the value dimensions that matter. My recommendation: start with five metrics per strategic supplier—two quantitative (cost and quality/delivery performance), two structured qualitative (innovation contribution and risk management effectiveness, assessed on a defined scale), and one open assessment (overall relationship health, bilateral). Review quarterly. Track them over time. Use the trends to identify relationships that are strengthening and relationships that need intervention. And use the data to make the case, internally and to the supplier, for continued investment in the partnership approach.

HYBRID-ERA SRM: HOW RELATIONSHIP MANAGEMENT ACTUALLY WORKS NOW

The way SRM gets executed day-to-day looks meaningfully different than it did a decade ago. Hybrid work (partly remote, partly in-person, with most strategic suppliers operating the same way) is now the operating environment, not a temporary adjustment. The practical question for procurement professionals isn't whether to manage relationships virtually. It's how to use virtual and in-person interaction in the right combinations.

Virtual interaction has clear strengths. Quarterly business reviews conducted by video conference make it trivial to include the right people: the engineering lead from the factory who has a quality question, the finance analyst who can address a billing dispute in real time, the executive sponsor who can participate in the strategic portion without the time commitment of traveling to a meeting site. The quality of supplier meetings often improves when the barrier to including relevant participants is low. Routine operational reviews, performance tracking sessions, and standard status updates all work well in virtual format and benefit from the efficiency.

In-person interaction does something different. The dinner after the QBR. The factory tour. The handshake. The hallway conversation where the supplier's account manager mentions something they wouldn't say in a formal meeting. These informal interactions build trust and generate intelligence that formal meetings don't, and they don't replicate well in video format.

The procurement teams who manage this tension most effectively are deliberate about which interactions happen virtually and which happen in person. Operational reviews, routine performance tracking, and standard updates work well virtually. Strategic alignment conversations, relationship-building sessions, and the kind of candid conversations that generate the deepest trust and the most valuable intelligence benefit from in-person interaction and are worth the travel investment.

My rule of thumb: for every strategic supplier, at least two in-person interactions per year—one at your site, one at theirs. Site visits are particularly valuable because they generate contextual understanding no virtual meeting can replicate. Walking through a supplier's factory, seeing their operations firsthand, meeting the people who actually do the work—that builds a relationship foundation months of video calls cannot.

SRM IN A DISRUPTED WORLD: RESILIENCE AS THE NEW KPI

Let me bring this chapter to its destination by connecting everything we've discussed to the reality procurement professionals are operating in today.

The world has not gotten simpler since the adversarial procurement model was born. It has gotten dramatically more complex. Supply chains have extended globally, creating efficiencies that have also created vulnerabilities. Geopolitical instability has become a persistent feature of the business environment rather than an occasional disruption. Climate change is generating supply shocks (floods, droughts, extreme weather events) with increasing frequency and unpredictability. Cybersecurity threats have made digital supply chains a vector for cascading risk. Regulatory environments are evolving rapidly across multiple dimensions (environmental, social, labor, data privacy) in ways that create compliance challenges extending deep into supplier networks.

In this environment, supply chain resilience, the ability to absorb disruptions and recover quickly, has become a strategic imperative that didn't appear on most procurement scorecards a generation ago. And supplier relationships are the single most important determinant of supply chain resilience. Not contracts. Not insurance. Not risk frameworks, though all of those matter. Relationships.

Because when the disruption hits, and it will hit, the question that determines how badly you suffer and how quickly you recover is not "what does our contract say?" It's "who picks up the phone at 11 PM when you call them?" It's "which customers does the supplier prioritize when allocating scarce supply?" It's "who are the twenty relationships across our supply base that we have invested in deeply enough that those people will move mountains to help us get through this?"

The COVID-19 pandemic was the ultimate test of this. The organizations with deep, trusting supplier partnerships discovered that those relationships were lifelines—suppliers who genuinely cared about their customers' survival worked extraordinary hours, found creative solutions, prioritized allocation in ways that kept businesses running. Suppliers who had been treated as adversaries found themselves with an opportunity to decide which customers were worth the extra effort, and they decided rationally based on the quality of the relationships they'd experienced.

The organizations that have built those partnership relationships, that have invested years in mutual trust and shared problem-solving, discover in moments of crisis that they have built something irreplaceable. The organizations that have spent those same years treating their suppliers as adversaries discover that their supply base has made a quiet, rational calculation about their relative importance —and acted accordingly.

BIG RON'S SUCCESSOR

Big Ron retires eventually. Everyone does. The chair he sat in for thirty years—the chair where he stared down vendor reps, where he ran competitions he sometimes manufactured for sport, where he extracted the last possible nickel of margin from people whose names he didn't bother to remember—gets filled by someone else.

His successor is in her thirties. She came up through procurement in the late 2010s. She runs Big Ron's portfolio —all of it, untouched since he handed it over. Her tools are unrecognizable.

What's on her screen every morning is a supplier health dashboard. It surfaces every supplier in her portfolio that has triggered an anomaly in the last twenty-four hours—not the routine noise, the signal. The parent company of a critical supplier just announced a leadership change in their commercial division. A second-tier supplier flagged margin pressure on their most recent earnings call, in the segment that touches her contract. A logistics partner has had three customer complaints surface on a niche industry forum in the last week. She knows all of it before her coffee finishes brewing. None of this required her to do any work. It required her to glance at the dashboard.

She doesn't do anything with most of it. Most of it is noise that resolves itself. But when a signal looks real, she does what Big Ron never had the time or the inclination to do: she calls the supplier. Not to ambush them. To check in. "Hey, I saw the news about your leadership change. Anything I should know about how it affects our work together?" That call, made early, surfaces information Big Ron would have learned about three months later, when his service levels mysteriously slipped and he started his accustomed cycle of escalation and pressure.

Her QBR prep takes ninety minutes instead of the two days Big Ron's analysts used to spend on it. Performance data synthesized automatically. The supplier's recent earnings, recent press, recent leadership changes mapped against the relationship's current pressure points. Likely issues the supplier will want to raise—anticipated. Likely issues she needs to raise—framed. She walks into the QBR with a position: here's where we are, here's where I think we're going, here's where I think you're going, let's compare notes. The supplier walks out feeling met. Big Ron's QBRs ended with both parties feeling vaguely dirty.

When a supplier brings her a problem, she runs it through her AI before she escalates. Not because the AI tells her what to do—it doesn't, and she wouldn't trust it if it tried—but because it gives her a fast synthesis of the supplier's situation, what happened when similar issues hit other companies, the likely root causes, and the range of options. By the time she's on the phone with the supplier's account team, she's thinking about the problem at the level of root cause and resolution. Big Ron, with the same call thirty years earlier, would have been thinking about leverage.

She has Big Ron's chair. The rest of the job is unrecognizable.

The point isn't that AI made her better than Big Ron. It didn't. She chose partnership over extraction before she ever logged into the dashboard. What AI did was make the partnership choice sustainable at the workload Big Ron's chair demands. The depth of supplier intelligence that partnership thrives on—knowing the supplier's pressures, anticipating their concerns, surfacing problems early—used to be impossible to maintain at scale. One analyst could do it for a handful of relationships. Not for the portfolio Big Ron ran. So most people, even those who'd have chosen partnership philosophically, ended up Big Ronning their way through the workload because there was no other way to cover it. That constraint is gone.

Same principle. New capability.

Now let's talk about where all of this gets tested. Where preparation meets pressure. Where everything we've built over years of careful relationship management gets put to use in a few hours of high-stakes conversation.

Let's talk about negotiation.

CHAPTER 2
THE ART OF THE DEAL

NEGOTIATION THAT ACTUALLY CREATES VALUE

Picture this scene.

It's 1994. A conference room somewhere in corporate America. Fluorescent lighting doing its level best to drain the joy out of everyone in the room. A procurement team on one side of the table. A supplier's sales team on the other. Coffee that's been sitting on the burner since seven AM. And a negotiation that is about to commence in the time-honored tradition that most procurement professionals of that era had been trained to execute.

The procurement lead opens with a number that is insultingly low. Not accidentally low—strategically, deliberately, almost offensively low. The kind of opening number that causes the supplier's account rep to do a quick mental calculation of whether their commission check is going to cover their mortgage. The supplier counters with a number that is equally absurd in the other direction. And then begins the ritual—the back and forth, the theatrical sighs, the "I need to check with my manager" gambits, the deadline pressure, the competitive threat that may or may not have any basis in reality, the final dramatic concession that both parties have been moving toward all along.

An hour and a half later, somebody shakes somebody's hand. The procurement team walks out having hit their savings target. The supplier's rep walks out having protected enough margin to make their quota. And absolutely nobody in that room has created a single dollar of new value. They have simply redistributed existing value (moved money from the supplier's pocket to the buyer's organization, or vice versa) through an elaborate performance that both sides have rehearsed so many times it has become almost entirely ritual.

This is what passed for negotiation excellence in the procurement world for most of the twentieth century. And while I am not going to stand here and tell you it never worked (it worked, in the narrow sense that it generated savings on paper) I am going to tell you that it left staggering amounts of value on the table every single time. Not the value you can see in a savings report. The invisible value: the innovation that never got proposed because the supplier knew you'd just use it as a price lever. The problem that never got solved collaboratively because the relationship wasn't built for collaboration. The risk that never got surfaced because transparency felt like vulnerability in an adversarial dynamic.

The negotiation that creates real, lasting, compounding value looks nothing like that scene. Let me show you.

THE TIME HEIST. 1994 EDITION.

The conference room hasn't changed since 1994. Same fluorescent lights, same procurement team across from the same supplier sales team, same coffee that's been on the burner since seven AM. What's changed is the preparation the procurement lead brought into the room.

Three days before the meeting, she ran a focused two-hour AI session on the supplier. The earnings calls were synthesized for margin pressure in the segment she was buying from. Recent customer wins and losses got mapped. Two analyst reports flagged that the supplier's regional sales VP had restructured the team six months ago—meaning the rep across the table from her was almost certainly under more pressure than a year prior.

She ran another prompt session mapping the negotiation's value dimensions. Not just price. Volume commitment, payment terms, contract length, exclusivity, performance guarantees, innovation provisions, risk allocation. Each one scored for how much it would matter to her organization versus the supplier. The insight: some of the supplier's high-value items were items her organization barely cared about. Longer term commitment in particular—she could give three years instead of one with no operational pain, and that single concession was likely worth meaningful per-unit price movement to a salesperson trying to lock in revenue certainty.

When the meeting started, she didn't open with an insulting-low number. She opened with a market-anchored proposal that her AI-assisted benchmarking had set fifteen percent below the supplier's published list price, supported by three competitor data points she could cite by name. The supplier's account rep recognized the work immediately. He couldn't dismiss her number as uninformed, and the ritual didn't begin. The actual conversation did.

Forty-five minutes in, she made an unexpected move. She didn't push on the price. She asked: "What's a three-year commitment worth to you?" The account rep, who had been bracing for a price fight, was thrown by the question. He had to think about it. He had to call his manager—for real this time, not as a gambit. When he came back, he had a number that was meaningfully better than what she could have squeezed out of him on price alone, and her organization had given up nothing of value to get it.

An hour later, somebody shook somebody's hand. The procurement team walked out having beaten their savings target by six percentage points. The supplier's rep walked out with a three-year revenue commitment that his VP would notice. And both parties had also surfaced a problem worth solving together: the supplier's onboarding process had been a chronic friction point for the buyer's stakeholders, and as it turned out, three other major customers had been raising the same complaint. Neither party would have raised it in the 1994 version of this meeting because the adversarial frame wouldn't have allowed it. In the 2026 version, it surfaced naturally—and became a joint initiative that delivered real operational value over the contract term.

This is the difference: the 1994 negotiation was a transaction; the 2026 version is a conversation about value. It's the same room, the same suppliers, the same product. What matters about the outcome—the preparation, the posture, the result—is different.

The procurement lead in this scene isn't smarter than her 1994 predecessor. She has better tools, and she's used them. The capability to map a supplier's pressures in two hours, to score the value dimensions against both parties' actual interests, to walk into the room with intelligence the supplier didn't expect—all of it sits a Tuesday afternoon away from any procurement organization willing to build the habit.

Same principle. New capability.

WHAT NEGOTIATION ACTUALLY IS (AND WHAT IT ISN'T)

Let's start with first principles, because the word "negotiation" carries so much cultural baggage that it's worth stripping it back to its actual meaning before we start adding layers of technique and framework on top.

Negotiation is the process by which two or more parties with different interests arrive at an agreement. That's it. That's the whole thing. At its core, negotiation is a communication process—a structured conversation about what each party needs, what each party can offer, and where the overlap between those needs and offerings creates the possibility of a mutually acceptable outcome.

Notice what's missing from that definition: winners and losers. The zero-sum framing (the idea that every dollar the buyer saves is a dollar the supplier loses, and that the goal is therefore to maximize your extraction at the other party's expense) is not inherent to negotiation. It's a choice. It's a model. And it's a model that consistently produces inferior outcomes compared to the alternatives.

Decades of academic and behavioral research have made a consistent and compelling case that the most effective negotiators are not the most aggressive ones. They are the most prepared—the ones who understand the other party's interests at least as well as their own, who can identify creative options that create value for both parties. Options that a purely positional, adversarial negotiator would never see because they're only looking at the price dimension of a multidimensional problem.

In procurement, this insight has profound practical implications. Most of what procurement professionals are negotiating is not, at its core, a simple price problem. It is a complex value problem—involving price, yes, but also quality, timing, risk, flexibility, service levels, innovation, relationship investment, and a dozen other dimensions that all have real value to one or both parties. The negotiator who understands all of those dimensions and can trade across them, giving up something of low value to one party in exchange for something of high value to the other, is creating value. Literally creating it, out of the difference between what each party values and what they can provide.

That is not the negotiation they taught in the 1994 conference room. But it is the negotiation that the best sourcing professionals in the profession have been quietly executing for decades. And it is the negotiation that the tools and intelligence now available are making more accessible to everyone with the preparation, data, and analytical power to practice it.

THE PREPARATION IMPERATIVE: WHY MOST NEGOTIATIONS ARE WON BEFORE THEY START

Here is the single most important thing I can tell you about negotiation: it is ninety percent preparation and ten percent execution. Maybe eighty-twenty on a good day. But the ratio is never remotely close to fifty-fifty, and any procurement professional who thinks the skill of negotiation lives primarily in the room (in the clever rejoinder, the well-timed silence, the dramatic walkout) has misunderstood where the real leverage comes from.

The negotiators who consistently win (who get better outcomes, more creative solutions, and more durable agreements than their peers) are relentless in their preparation. They know things walking into the room that the other party doesn't expect them to know, have already mapped the range of possible outcomes that create the most value for their organization, and have thought through the other side's interests and constraints so carefully they can practically predict the other side's moves before they happen. And they have identified their alternatives so clearly that they know when to walk away—which means they are never trapped by the fear of losing a deal that isn't actually in their best interest.

Let me walk you through the preparation framework that has served me across thirty years of procurement negotiations—from modest contract renewals to nine-figure strategic sourcing programs.

Know Your Anchors: BATNA and Reservation Point

Before you walk into any negotiation, you need to know two numbers with absolute clarity. The first is your BATNA—your Best Alternative To a Negotiated Agreement. This is what happens if you don't reach a deal with this supplier. It might be awarding to a different supplier. It might be insourcing the capability. It might be doing without whatever you were trying to buy. The BATNA is critical because it defines your real leverage. A strong BATNA, a real alternative you could actually execute, gives you the freedom to walk away from a bad deal and the confidence to push hard in the negotiation. A weak BATNA (no viable alternative, high switching costs, critical need that can only be met by this supplier) limits your leverage and requires you to compensate with preparation and creativity.

The second number is your reservation point—the worst outcome you would accept before walking away. Below your reservation point, you're better off exercising your BATNA than accepting the deal on offer. Knowing this number precisely, before you're in the room, prevents you from making concessions under pressure that you'll regret afterward.

Here's the critical insight that most of the best buyers

miss: the BATNA isn't something that just exists in the market. It's something you can build. Procurement professionals who invest in developing real alternatives (qualifying backup suppliers, building internal capabilities, maintaining competitive pressure in categories where they've been sole-sourced) are systematically improving their negotiating position across their entire portfolio. This is strategic, proactive procurement at its best.

Know Their Anchors: The Other Side's BATNA and Interests

Knowing your own position is necessary but not sufficient. The preparation that separates great negotiators from good ones is the depth of their understanding of the other party's situation.

What is the supplier's BATNA? What happens to them if they don't get this deal? Are they at quota? Are they fighting to hold a key account for reference customer purposes? Are they trying to break into your industry and willing to make significant concessions to do so? Are they in a financially stressed position that makes near-term revenue more valuable than it would otherwise be? Every piece of intelligence about the supplier's situation is potential leverage—or, equally valuable, a signal about where there is room to create mutual value rather than just extract it.

Understanding the supplier's interests (not just their stated position, but the underlying business objectives and constraints that drive their position) is where the real magic of principled negotiation happens. A supplier who opens with a high price isn't just being greedy. They might be managing margin pressure from their own raw material costs, protecting a price point that matters for their broader market positioning, or pricing risk into their number because they're skeptical about the volume commitments you're making. Understanding which of these dynamics is actually at play allows you to address the real issue rather than just haggling about the number.

How do you get this intelligence? Research, primarily. Annual reports and financial filings for public companies. Industry analyst reports. LinkedIn intelligence about who's on the account team and what their incentive structure looks like. Conversations with references and other customers. Your own history with the supplier, if it exists. And increasingly, AI-powered market intelligence tools that can synthesize enormous amounts of publicly available information into supplier-specific insight in a fraction of the time traditional research required.

Let me give you a concrete example of what this can produce. Last year, I prepared for a renewal with a major logistics provider—a mid-seven-figure contract, three-year term coming up. In the old world, thorough preparation would have taken my team two weeks of pulling filings, reading analyst reports, benchmarking pricing, and assembling a picture of the supplier's negotiating pressures.

I did most of the same work in a focused ninety-minute session with a generative AI tool. Earnings call transcripts synthesized for margin pressure, pricing strategy, and customer retention signals. Public financials flagged for the trends that would matter—declining ground freight margins, heavy capex on warehouse automation that needed volume to amortize, an executive restructuring in their enterprise sales organization. Published rate structures benchmarked against three competitors in our geography.

The insight that changed the negotiation came from synthesis. The AI surfaced that the supplier had recently lost two major accounts in our region within the same quarter—information scattered across a local trade publication, a LinkedIn post from a departing account manager, and a brief mention in an analyst Q&A. None of those data points alone would have caught my attention. Synthesized, they painted a clear picture: this supplier had excess capacity in our geography and significantly more motivation to retain our business than their confident opening position suggested.

That single insight, assembled from publicly available information, shifted the negotiation. It didn't replace judgment, relationship management, or tactical execution. But it gave me a preparation advantage that, five years ago, would have required a research team and weeks of calendar time.

Map the Value Dimensions

Most negotiations get treated as one-dimensional—price. But almost every meaningful procurement negotiation is actually multidimensional, and the procurement professional who maps all the dimensions before walking into the room has options that the one-dimensional negotiator doesn't.

What are the value dimensions in a typical procurement negotiation? Price, obviously—the per-unit or total contract price. Payment terms—how quickly you pay, which has real cash flow value for both parties. Volume commitments—what you're promising to buy, which affects the supplier's capacity planning and risk. Contract length—longer terms provide security for both parties and often unlock better economics. Performance standards and penalties—how you define and enforce quality and service levels. Exclusivity provisions—whether you're committing to a single source. Innovation and development commitments—whether either party is investing in joint development of new capabilities. Risk allocation—who bears the cost if things go wrong in various scenarios.

Each of these dimensions has different value to each party. Payment terms might be low value to you but high value to a supplier managing cash flow. Volume commitments might be high value to you (you want flexibility) and high value to the supplier too (they want predictability). Contract length might be something you'd trade away easily, you don't love long commitments, but something the supplier would pay significantly for in commercial terms because it reduces their sales cost and provides revenue certainty.

When you map these dimensions carefully before the negotiation, you can identify trades (give something of low value to you in exchange for something of high value to you, and find the symmetric trade from the supplier's side) that create real value a purely price-focused negotiation would never surface. Fisher and Ury called this "expanding the pie before dividing it." And it is, consistently, where the most sophisticated procurement negotiations find their best outcomes.

THE NEGOTIATION TABLE: TACTICS, PSYCHOLOGY, AND THE HUMAN ELEMENT

Mike Tyson said it best. Everyone has a plan until they get punched in the face.

My punch came from a guy named Chad.

I'm not going to give you the details, anonymity protects the guilty, but I will tell you that I walked into that conference room as prepared as I had ever been for a negotiation in my career. Market data. Alternatives qualified. Value dimensions mapped. Internal alignment locked down. I had a plan. I had a backup plan. I had a backup plan for the backup plan. And forty-five seconds in, Chad opened with a move I had not anticipated and was, frankly, not prepared to handle. My plan did not survive contact with Chad.

(For the record: I recovered. We got the deal done. But there was a hallway recess in there that I don't talk about at parties.)

Here's the thing about the punch in the face. Tyson wasn't saying preparation doesn't matter. He was saying preparation isn't the whole job. Preparation is what gets you into the ring. What happens once the bell rings (once a Chad rolls out a tactic you didn't expect, once the stakeholder you trusted decides to undermine you, once the supplier proposes a pricing structure that doesn't fit any of the frameworks you researched) is execution. And execution is its own discipline.

Preparation wins negotiations. But you still have to execute. And execution involves navigating the human psychology of the negotiation table—the dynamics, the tactics, the moments of tension and creativity that transform preparation into outcomes.

Let me share the most important principles I've accumulated across three decades of sitting at those tables. Including the ones I learned the hard way, courtesy of every Chad I've ever encountered.

Anchor First, Anchor Boldly

The first number put on the table in a negotiation has a disproportionate influence on where the negotiation ends up. This is one of the most replicated findings in behavioral economics, the anchoring effect, and it is as real and powerful in procurement negotiations as in any other context.

The practical implication: if you have a strong BATNA and good market intelligence, anchor aggressively. Not insultingly—the goal is not to offend the other party or signal that you haven't done your homework. But make your first position ambitious enough that the negotiation moves in your direction rather than theirs. The procurement pro who responds to the supplier's opening with a mild counter has already ceded significant ground. The sourcing professional who opens with a well-researched, confident anchor sets the gravitational center of the negotiation much closer to their desired outcome.

What makes an anchor credible rather than merely aggressive? Market data. The anchor that says "our research shows market pricing for this category is X, which is where we're starting" is far more powerful than an arbitrary low-ball that the supplier can dismiss as uninformed. This is another place where preparation, and specifically market intelligence preparation, translates directly into negotiation leverage.

The Power of Silence

Amateur negotiators cannot tolerate silence. The moment a pause opens up in a negotiation (after they've made a proposal, after they've pushed back on a supplier's position, after they've made a concession) they fill it. With explanation. With justification. With additional concessions they hadn't planned to make. With words that dilute the impact of the position they just staked out.

Professional negotiators have learned, usually through uncomfortable experience, that silence is one of the most powerful tools at the table. When you make a proposal and then go quiet, you are putting the full weight of the moment on the other party. You are signaling confidence in your position. And you are creating the conditions for the other party to fill the silence—often by making concessions or revealing constraints that move the negotiation in your direction.

The discomfort of silence in a negotiation is real and visceral. Learning to sit in it, to let it work for you rather than rushing to relieve it, is a skill that takes practice and self-awareness. Some of the most significant concessions I've ever received in negotiations came from simply shutting up after making a demand and waiting.

Make Concessions Strategically

Not all concessions are created equal. How you make concessions (the size, the timing, the framing, and the conditions) communicates information about your position and shapes the other party's expectations about what additional concessions might be available.

The classic mistake is making large, rapid concessions in response to pressure. This communicates two things simultaneously: that your opening position wasn't credible, and that pressure works. Both of those signals are catastrophic for your negotiating position going forward. The supplier who learns that pushing back hard produces rapid large concessions will push back hard early and often in every negotiation.

Strategic concession-making involves making concessions that are progressively smaller, signaling you're approaching your limit, and always conditional. "If you can move on the payment terms, I can revisit the volume commitment" is a conditional concession that trades value rather than simply surrendering it. It keeps the negotiation multidimensional and creates a path toward the mutually beneficial outcome rather than simply transferring value from one side to the other.

Make concessions slowly. Make them smaller as you go. Make them conditional whenever possible. And make them feel earned—because concessions that feel earned generate reciprocity, while concessions that feel surrendered generate more demands.

Managing the Emotional Register

Negotiation is an inherently emotional experience, regardless of how much both parties might prefer to think of it as a purely rational exercise. Deals feel personal; concessions feel like defeats; pressure tactics feel like disrespect. And the emotional temperature of a negotiation has enormous influence on its outcomes.

The procurement teams who manage the emotional register most effectively are the ones who are secure in their preparation. When you know your BATNA, when you understand the market, when you've mapped the value dimensions, you are not emotionally dependent on the outcome of this particular negotiation. That security shows. It is the source of the calm confidence the best negotiators project—not a practiced persona, but the natural expression of someone who has done the work and knows where they stand.

Managing the other party's emotional state is equally important. Skilled negotiators recognize when the other side is feeling cornered, when the pressure they're applying is generating defensiveness rather than movement, and they know how to release that pressure without surrendering commercial ground. A well-timed recess. A pivot from competitive to collaborative framing. An acknowledgment of the other party's constraints that signals you understand their situation even as you're pressing your own. These are techniques that require emotional intelligence and situational awareness no framework can fully codify—they come from experience, from paying attention, and from caring about outcomes rather than just performance.

THE PRINCIPLED NEGOTIATION FRAMEWORK: INTEREST-BASED BARGAINING IN PRACTICE

Fisher and Ury's "Getting to Yes," published in 1981 and still one of the most important books on negotiation ever written, introduced a framework called principled negotiation or interest-based bargaining that has become foundational to professional negotiation practice. Its core insight is deceptively simple: the positions people state in a negotiation are not the same as the interests underlying those positions, and the path to better outcomes runs through understanding interests rather than fighting over positions.

Here's a classic example from procurement. A supplier states their position: the price is $100 per unit and they can't go below it. Your position: you need to pay $80 per unit or the deal doesn't work. You're $20 apart. If you negotiate purely on positions, you'll probably split the difference somewhere ($90, perhaps) and both parties will leave the table feeling like they lost something without gaining much of anything.

But what if you explored the interests behind those positions? The supplier's interest might be protecting a gross margin their CFO has set as a minimum threshold for new contracts. Your interest might be hitting a total cost target for a product launch with a fixed budget. Those interests might be reconcilable in ways a pure price negotiation would never discover. Perhaps a longer contract term that gives the supplier revenue certainty lets them reduce the per-unit price without violating their margin floor. Or a higher initial volume commitment—something you could actually offer given your launch forecast—that changes the supplier's unit economics enough to bridge the gap. A payment-terms improvement might be worth something commercially that allows real price movement.

None of those options are visible if you're only looking at the price number. They become visible the moment you start asking "why is $100 their floor?" and "what would need to be true for them to move?" That question, the curiosity about what's driving the other party's position, is the heart of principled negotiation.

In practice, interest-based bargaining in procurement involves a set of behaviors that feel counterintuitive to procurement professionals trained in the adversarial school. It involves asking questions early and often—not leading questions designed to entrap, but inquiries aimed at understanding. It involves sharing information strategically—not revealing your reservation point, but being transparent about your interests and constraints in ways that open up the solution space. It involves being explicit about the value dimensions you're willing to trade across, so the other party understands you're not just trying to extract on price but are looking for the deal structure that works for both parties.

And it involves being willing to be creative. The best

procurement negotiations I've ever been part of, the ones that created the most value for my organization while building or maintaining the supplier relationships that delivered long-term competitive advantage, were not won through superior toughness. They were won through superior creativity. Through finding the option that neither party had considered at the outset, that solved both parties' underlying problems, and that turned what looked like an impasse into an agreement both sides were pleased with.

That's not a soft outcome. That's a superior outcome. And it's the outcome that principled, interest-based negotiation is designed to produce.

CATEGORY-SPECIFIC NEGOTIATION: ONE SIZE DOES NOT FIT ALL

One of the most important things I learned over the course of my career, and one of the things that took me longer to internalize than I'd like to admit, is that negotiation strategy has to be calibrated to the specific category and relationship context you're operating in. The approach that works brilliantly in a commodity category with abundant supply and multiple qualified suppliers will undermine you in a specialized technology category where you're deeply dependent on a single provider's unique capabilities.

Let me walk through the key category contexts and how negotiation strategy should adapt to each.

Commodity Categories

In truly commoditized categories (where the product or service is largely undifferentiated, where multiple qualified suppliers exist, where switching costs are low) the negotiation dynamic is closest to what the traditional procurement model was designed for. Competitive tension is real and powerful, market benchmarks are meaningful, and price-focused negotiation delivers genuine value. This is the category context where reverse auctions, aggressive anchoring, and explicit competitive pressure are all appropriate tools.

Even here, though, the best procurement pros layer in more sophistication than pure price competition. Total cost of ownership (including logistics, quality, administrative burden, and supplier reliability) matters even in commodity categories. The supplier who is slightly more expensive on unit price but dramatically more reliable on delivery and quality frequently delivers better total value than the cheapest bidder.

Strategic Differentiated Categories

Most procurement work happens in the middle. Categories where multiple qualified suppliers exist, so competitive tension is real, but where suppliers are meaningfully differentiated in capability, quality, service model, or strategic fit. Indirect spend like marketing services, engineering consulting, manufactured components with technical specifications, mid-tier technology platforms. The supplier you choose matters; you have credible options; switching is feasible but not trivial.

The negotiation challenge in this category is that pure price competition leaves value on the table (because the suppliers are not, in fact, interchangeable) but pure relationship-driven negotiation gives up the leverage your competitive options provide. The sourcing professionals who get the best outcomes here run a hybrid play: rigorous competitive process at the front end of selection, principled value-based negotiation once the field is narrowed, and a deliberate decision about the relationship investment level once the supplier is selected.

The specific moves matter. Run the competition on a defined scope—make sure you're comparing comparable offerings, not apples and oranges. Use the competitive intelligence to inform the negotiation, but don't anchor the entire negotiation on price alone—the value dimensions framework is most useful in this category, because there are real trade-offs available. Once you've selected, treat the relationship as the strategic asset it is—the differentiation that drove your selection is also the differentiation that creates switching cost over time, and the relationship investment compounds. The supplier who feels chosen on merit and treated as a partner from day one will perform differently from the one who feels squeezed through the selection process and then dumped into transactional management.

This is the category where the disciplines we've discussed in this chapter and the partnership disciplines from Chapter One come together most explicitly. Get this category right and you've internalized the substance of the first three chapters.

Strategic Sole-Source Categories

At the opposite end of the spectrum are strategic sole-source categories—where you're dependent on a single supplier with unique capabilities you cannot readily replace. This is the category context where the adversarial model is not just suboptimal but potentially catastrophic. Beating up a supplier you can't replace doesn't save money. It damages the relationship, degrades service quality, and destroys the supplier's investment in your account—all without reducing their ability to extract pricing power from you at the next renewal.

Negotiation in strategic sole-source situations requires a completely different toolkit. Preparation focuses on deep understanding of the supplier's cost structure and economics—not to use as a hammer, but to identify where genuine cost reduction is possible through collaboration versus where pushing on price is simply transferring margin at the cost of service quality. It requires building the relationship capital that creates flexibility—because in a sole-source situation, your leverage is not competitive threat (which isn't credible) but the value of the ongoing partnership to the supplier. And it requires a multi-year perspective—managing the commercial terms across the full strategic relationship arc rather than optimizing each individual contract renewal in isolation.

Technology and Services Categories

Technology and professional services occupy a distinctive middle ground—often with multiple suppliers at the outset of a sourcing process, but with significant switching costs once a supplier is embedded. This dynamic creates a temporal negotiation challenge: your leverage is highest before you select a supplier and commit to implementation, and it declines dramatically after go-live when the cost and disruption of switching become real.

The procurement professionals who consistently get the best outcomes in technology and services categories are the ones who do the hard negotiation work before selection—not after. They negotiate implementation costs, transition assistance, price escalation caps, performance guarantees, and exit provisions while they still have credible competitive alternatives. They build detailed contractual protections against the most common value erosion mechanisms in these categories: scope creep, price increases at renewal, support tier degradation, and the increasingly common practice of vendors acquiring complementary companies and then forcing customers onto consolidated, and usually more expensive, combined platforms.

The Year Three Problem

There's a specific dynamic in technology and services contracts worth naming explicitly, because it explains why so many of these relationships go sideways at renewal: the year three problem. Year one, you negotiate hard because you have competitive options and the supplier is trying to win the deal. Year two, you implement and integrate; the supplier is investing to make you successful, and the relationship feels good. Year three, the initial contract term expires—your switching costs have fully materialized, and the supplier knows exactly how captive you are. And the renewal negotiation that you thought was about market pricing turns out to be about how much margin the supplier can extract from a customer who can't credibly walk away.

The procurement professional who has not prepared for year three during the year one negotiation will have a very bad year three. The one who builds the right contractual protections at the front end will find year three looks remarkably like year one. The protections worth fighting for: price escalation caps that limit annual increases to a defined formula (not "market" pricing—"market" is whatever the supplier says it is at renewal); explicit exit provisions covering data portability, transition assistance, and termination support; audit rights on usage and pricing; performance guarantees with real consequences for sustained underperformance; and renewal terms that prevent the supplier from forcing acceptance of commercially unfavorable changes as a condition of continued service.

During years one and two, do the work that keeps your year-three options live. Maintain ongoing competitive intelligence—at least annually, refresh your view of who else is in the market and what their offerings look like. Where it's economically viable, run small qualifying engagements with potential alternatives so you have a credible, partially proven option if you need to switch. Track your supplier's market position (acquisitions, leadership changes, financial pressure) for signals about how their priorities and pricing posture might shift. The goal is not to be ready to switch at year three; it's to be credibly able to switch, which is what gives you negotiating leverage even if you ultimately renew with the incumbent.

This work is not glamorous and not directly tied to current-year savings targets. It's the procurement equivalent of building optionality. And it's one of the highest-leverage things a category manager can do in any technology or services category.

THE INTERNAL NEGOTIATION: WHY YOUR MOST POWERFUL ALLY ISN'T AT THE TABLE

Everything we've discussed so far in this chapter has been about the negotiation between you and the supplier. But some of the most consequential negotiation work a procurement professional does happens before the supplier is ever in the room—and it happens with your own stakeholders.

Let me tell you about one of the most effective negotiations I've ever been part of—and the move that made it work had nothing to do with market data, BATNA analysis, or any of the frameworks we've covered. It had to do with a business stakeholder who was willing to trust procurement enough to run a play together.

The situation was a textbook procurement nightmare. We were coming up on a seven-figure annual renewal with a dominant cloud platform vendor in the life sciences space—the kind of vendor that owns their vertical so completely that the switching costs aren't just high, they're essentially prohibitive. Their software was embedded in our regulated processes. There was no credible alternative we could have migrated to in any reasonable timeframe. On paper, our BATNA was nonexistent. We were, by any structural analysis, captive.

The vendor knew this. Their sales team had a playbook for customers in our position, and I had seen it executed enough times to know exactly what was coming: the account rep and their regional VP would walk us through a story about increased platform investment, expanded R&D costs, new modules in the pipeline, and the value of the "partnership"—all building toward a price increase that they would frame as modest and inevitable. The goal of their playbook was to move us off our cost optimization posture and into a defensive position where we were negotiating the size of the increase rather than pushing for a decrease.

Here's where the internal negotiation became critical. I had a pre-meeting with my primary business stakeholder—the executive who owned the platform from the user side. I walked him through exactly what the vendor was going to say in our upcoming session. The talking points, the framing, the arguments they'd use to justify the increase. I told him how they'd try to anchor the conversation around their investment narrative rather than our value realization.

He sat through my briefing, and when I finished, he looked at me and said: "Wow. You were right about everything. So what do you want to do?"

That moment, the moment where a stakeholder realizes that procurement actually understands the vendor's playbook better than they do, is where real internal trust gets built. And that trust is what made the next part possible.

I suggested we run a coordinated play. Good cop, bad cop—but with a twist that made it far more effective than the standard version.

When the actual negotiation session came, my stakeholder sat on the supplier's side of the table. Not next to me. Across from me, physically aligned with the vendor's account rep and their regional VP. This was deliberate. I wanted the supplier's team to feel, subconsciously, that they had an ally in the room—that the business user was on their side and the procurement guy was the obstacle. It shifted their mental model of the room's dynamics in a way that would matter enormously in about forty-five minutes.

The meeting proceeded exactly as I'd predicted. The regional VP set expectations, the account rep walked through the value story, and as they made their case for why pricing needed to move in their direction, my stakeholder listened—and then, with a conviction that I had not fully anticipated, started pushing back. Hard.

He took umbrage at specific claims they made. Not performed umbrage—this looked and felt real. The room got tense. The kind of tense where people stop looking at their laptops and start watching body language. And then, after a particularly pointed exchange, my stakeholder stood up and walked out of the room.

The silence that followed was extraordinary.

Remember—this wasn't the procurement guy walking

out. Procurement walkouts are a known tactic; sophisticated sellers expect them and discount them accordingly. This was the business owner. The person who actually uses the platform every day. The person whose renewal signature the vendor needed. And he had just left the room visibly angry, with the regional VP sitting across the table trying to figure out whether the deal they'd been counting on was about to disappear.

I reacted like I was stunned. Because in the story we were telling, I was supposed to be stunned—the procurement professional who had been trying to find middle ground, now watching his own stakeholder blow up the conversation. I excused myself to go find him, to try to calm him down. I left the room. He was, of course, in a different conference room down the hall drinking coffee. I waited a few minutes, walked back in, and told them I couldn't find him.

The body language in that room when I came back was completely different from when I'd left. The regional VP who had walked in projecting confidence and control was now leaning forward, looking concerned. The account rep was making notes. The dynamic had shifted from "how do we get this customer to accept our increase" to "how do we save this deal."

I played the role of the reasonable person trying to hold things together. I told them I wanted to continue the conversation, that I thought we could find a path forward, but that my stakeholder was seriously questioning whether this was a partnership worth continuing—and that I needed them to help me make the case for renewal by giving me something meaningful to bring back to him.

The result: fifteen percent off the renewal price on a seven-figure annual contract over three years. Two new software modules, products that were on their roadmap and that we would have been charged separately for, included at no additional cost. And payment terms improved from net thirty to net sixty, which had real cash flow value for our organization.

All of this from a negotiation where our structural leverage was, on paper, zero.

The lessons here go beyond the specific tactic. The play only worked because I had already built credibility with the stakeholder. My ability to predict the vendor's playbook before they ran it (to walk him through the talking points, the framing, the arguments they'd use) transformed me, in his mind, from "the procurement guy who slows things down" into "the person who actually understands how these vendors operate." That credibility wasn't conjured for the meeting. It was earned over months of working together, of me delivering value in previous engagements, of him seeing that procurement's involvement made his outcomes better rather than just slower. Without that foundation, the same tactic in the same room would have produced a different result.

The play was also effective in a way that depended specifically on the angry party being the stakeholder, not procurement. When the business owner threatens to walk away, it carries a weight that a procurement walkout never can—because the vendor knows that if the user doesn't want the product, no amount of procurement negotiation is going to save the deal. The stakeholder's dissatisfaction was, in the vendor's eyes, an existential threat to the account. Mine would have been a tactic. The difference is everything.

The walkout was maximally destabilizing precisely because of where the stakeholder was sitting. The vendor's team had been reading the room as two-against-one in their favor. When their perceived ally turned hostile, the disorientation was real.

And underneath all of it: this kind of coordinated play requires trust between procurement and the business that no pre-meeting can manufacture. My stakeholder had to trust the strategy was sound. I had to trust he could execute the role. We had to be aligned on objectives, on limits, on exactly how far to push. None of that gets built in the briefing where you plan the tactic. It gets built over months and years of small choices, of procurement showing up as a strategic partner rather than a process obstacle, of every previous interaction earning the right to ask for trust at the moment it matters most.

The internal negotiation (the work of building stakeholder trust, earning credibility with the business, and creating the conditions for coordinated strategy) is, I would argue, as important as anything that happens at the supplier table. The procurement professional who has the business's trust can do things that the procurement professional who is merely tolerated cannot. And in situations where structural leverage is limited (which, in the age of dominant SaaS platforms and high-switching-cost technology categories, is increasingly common) stakeholder alignment may be the most powerful leverage you have.

NEGOTIATION ETHICS: THE LINES THAT SHOULD NEVER BE CROSSED

I would be doing this chapter a disservice if I didn't address something that tends to get glossed over in negotiation training but that matters enormously in professional practice: ethics.

Procurement negotiations involve real pressure, real stakes, and real temptations to cut corners on professional ethics in pursuit of a better commercial outcome. And while I've spent the last several thousand words arguing that the adversarial model produces inferior outcomes and that principled negotiation creates superior value, I want to be explicit that the case for ethical negotiation is not purely pragmatic. It's also, simply and fundamentally, the right way to practice this profession.

Some practices are clearly out of bounds. Misrepresenting your competitive position (fabricating bids, inventing alternatives that don't exist, lying about what competing suppliers have offered) is deceptive and unethical, and it will destroy your credibility and your organization's reputation when the truth emerges, as it usually does. Using confidential information that a supplier has shared in trust (their cost breakdowns, their technology roadmap, their pricing strategy) to benefit a competing supplier is a breach of professional ethics and in some contexts a legal violation. Conducting competitive sourcing processes that are predetermined, where the winner is already chosen and the competition is theater, wastes everyone's time, destroys your credibility with the supply market, and is a betrayal of the fairness principles that make competitive sourcing legitimate.

There are harder cases too—the gray zones where reasonable buyers can disagree. How much information are you obligated to share with a supplier about your alternatives? How aggressive an anchor is too aggressive? When does strategic ambiguity cross the line into misrepresentation? These questions don't have simple universal answers. They require professional judgment, grounded in a commitment to honesty and fairness, applied thoughtfully to the specific facts of each situation.

What I can say with confidence is this: the procurement professionals who develop reputations for fair, honest, professional negotiation (who are tough but trustworthy, demanding but respectful, disciplined but transparent) build something over the course of their careers that has real, compounding value. Their word means something. Their processes are taken seriously by suppliers who believe the competition is real, and their relationships survive the inevitable tough moments because the foundation of trust has been built and maintained.

Reputation, in procurement negotiation, is not soft. It is leverage.

THE MULTI-ROUND NEGOTIATION: PLAYING THE LONG GAME

Most negotiation training, and most of what we've covered so far in this chapter, focuses on the single session: one meeting, one set of issues, one outcome. But many of the most important procurement negotiations actually unfold across multiple rounds, sometimes over weeks or months, with the dynamics evolving between sessions in ways that create both opportunities and risks.

The multi-round negotiation is the norm in complex procurement. A major technology sourcing program might involve an initial requirements alignment session, followed by supplier presentations, a technical deep-dive, commercial discussions, a best-and-final-offer round, and finally a contract negotiation that itself might take several sessions. Each round has its own dynamics, its own objectives, and its own tactical considerations. And the way you handle each round affects the dynamics of every subsequent round.

The principles I've found most important in multi-round negotiations:

Control the pace.

In a multi-round negotiation, time is a strategic resource. The party that controls the pace—that sets the meeting schedule, defines the agenda for each round, and determines when the process moves from one phase to the next—has a significant structural advantage. They can accelerate when momentum is in their favor and slow down when they need time to regroup or develop alternatives. Procurement professionals should own the process design for any multi-round negotiation, and they should use that process ownership to create the time and space they need for thorough preparation between rounds.

Track concessions across rounds,

not just within them.

A common mistake in multi-round negotiations is losing track of the cumulative concession pattern across the full negotiation arc. The supplier who makes a significant concession in round two may try to claw it back in round four through a seemingly unrelated term. Maintaining a detailed concession log—what was proposed, what was conceded, what was conditioned—across all rounds is essential for ensuring that value gained in one session isn't quietly surrendered in another.

Use the gaps between rounds strategically.

The time between negotiation sessions is not dead time. It's preparation time. After each round, debrief immediately while the conversation is fresh. What did you learn about the supplier's priorities that you didn't know before? What arguments resonated and which fell flat? What's the supplier's body language telling you about their reservation point? Use the time to refresh your intelligence—particularly if the supplier made claims about market conditions or competitive dynamics you can verify or challenge in the next session.

Don't reveal your full strategy in round one.

Multi-round negotiations reward the patient procurement professional who reveals positions, options, and information strategically across sessions rather than putting everything on the table immediately. In round one, establish the framework and understand the supplier's position. In round two, introduce your market intelligence and value dimension analysis. In round three, begin proposing creative options. This sequencing keeps the supplier engaged and responsive throughout the process rather than feeling like they've been hit with everything at once.

Be willing to expand the scope between rounds.

One of the advantages of a multi-round format is that it gives both parties time to consider possibilities that weren't on the table initially. The creative option that seemed impractical in round one might become feasible after both parties have had time to think about it, discuss it internally, and develop the details. The best multi-round negotiations produce outcomes that neither party envisioned at the outset—because the extended process gave the creativity time to develop.

THE TEAM NEGOTIATION: ROLES, CHOREOGRAPHY, AND COORDINATION

Most significant procurement negotiations involve multiple people on the buying side—the category manager, perhaps a procurement director, often a business stakeholder, sometimes a legal representative, and occasionally a technical expert. Coordinating this team is itself a negotiation skill that gets surprisingly little attention in the training literature.

The most important principle of team negotiation is role clarity. Before the meeting, every person on your side of the table should know: what is their role (lead negotiator, subject matter expert, observer, decision-maker), when are they expected to speak and on what topics, what are they authorized to commit to, and what requires them to defer to someone else.

The lead negotiator should be clearly identified and should be the primary voice for the buying team. Multiple people from the same side making proposals, asking questions, and offering concessions creates confusion and undermines negotiating discipline. Others on the team should contribute in their areas of expertise (the technical expert clarifies requirements, the legal representative addresses contract terms, the stakeholder speaks to business impact) but the commercial positions and the negotiation strategy should flow through a single voice.

Pre-meeting choreography matters more than most procurement teams realize. Brief your team not just on the strategy but on the specific plays you might run. "If they push back on price, I'm going to pivot to the value dimensions. When I do, Sarah, that's your cue to talk about the quality issues we've been experiencing—I want them to feel the operational pain before we come back to price." That kind of coordinated team play—exactly what the stakeholder story earlier in this chapter illustrated—requires advance planning, trust among team members, and the discipline to execute roles under the pressure of a live negotiation.

One more principle, learned through painful experience: debrief as a team after every session, and do it immediately. Not the next day. Not next week. Now. The insights, observations, and instincts that team members have right after a negotiation are perishable. The supplier's body language when you proposed the pricing restructure. The account manager's reaction when you mentioned the competitive alternative. The moment when the room's energy shifted from defensive to collaborative. These observations are extraordinarily valuable for preparing for the next round, and they evaporate within hours if they're not captured while they're still fresh.

The team negotiation (properly prepared, properly choreographed, properly debriefed) is preparation discipline made visible. Everything we discussed in the preparation section earlier in this chapter applies, just multiplied across more people. The teams that practice it consistently are the ones that turn structural disadvantages into outcomes that look impossible from the outside.

THE NEGOTIATION MINDSET: WHAT SEPARATES THE GREAT ONES

I want to close this chapter the way I close every negotiation training I deliver—not with a tactic or a framework, but with a mindset observation.

The procurement professionals who are the best negotiators I've ever encountered share something that goes beyond preparation discipline and tactical sophistication. They are curious about the other party—they find the supplier's business interesting and want to understand how it works, what pressures it faces, what it's trying to accomplish. And that curiosity, that real interest in the person and organization across the table, serves them in ways no amount of tactical training can replicate.

It serves them because people can tell the difference between curiosity and strategic information extraction, and they respond to each very differently. The supplier account team that feels understood, that believes you actually grasp their situation, engages with you in a fundamentally different way than the one that feels interrogated. They problem-solve with you rather than against you, share information they might otherwise withhold, and advocate for you internally with their own leadership when you need flexibility they don't have authority to grant themselves.

Curiosity also serves the great negotiators because it keeps them learning. The procurement professional who is always trying to understand how supplier businesses work, how markets function, how costs are built up and margins are calculated, is always building the knowledge base that makes their preparation more accurate and their insights more valuable. They are, over time, becoming more dangerous in negotiations—not because they're getting tougher, but because they're getting smarter.

There's a version of the procurement profession that is fundamentally about transactions—buying things, hitting savings targets, processing purchase orders, checking the boxes on a sourcing methodology. That version of the profession is being automated.

The version that endures (the version that creates lasting, compounding value for organizations) is fundamentally about relationships, judgment, and the intelligence to create outcomes that wouldn't exist without skilled human guidance.

Negotiation, at its best, is the concentrated expression of all of that. It's where the thirty years of experience in the room translate into the few hours of conversation that determine whether an organization pays too much for too little, or gets exactly what it needs at a price that works, with a supplier invested in making the relationship succeed.

Get good at that, and you've gotten good at something that matters.

PART TWO

AI TRANSFORMS THE PRACTICE

HOW AI CAPABILITIES RESHAPE EVERY WORKFLOW THAT DEFINES THE MODERN PROCUREMENT FUNCTION.

CHAPTER 3
THE MACHINE IN THE ROOM

WHAT PROCUREMENT PROFESSIONALS ACTUALLY NEED TO KNOW ABOUT AI

Let me tell you about the moment I realized the game had changed.

It was a Tuesday afternoon in late 2023. I was sitting in my home office, staring at a sourcing project that had been slowly eating my calendar for the better part of two weeks. We were evaluating a new managed services provider for a mid-eight-figure outsourcing engagement—the kind of deal where getting it wrong doesn't just cost money, it costs careers. I had three finalist suppliers. I had a hundred-page RFP response from each of them. I had financial statements, reference check notes, analyst reports, and a contract redline from legal that looked like someone had attacked it with a red crayon. And I had about four days before the steering committee presentation where I was supposed to walk a room full of executives through a recommendation with the kind of confidence that makes people nod rather than ask uncomfortable follow-up questions.

In the old world, the world I'd been operating in for three decades, this was a two-week sprint, minimum. You'd divide the RFP responses among your team—somebody on the scoring matrix, somebody else on the financial comparisons. You'd spend hours reading dense prose about "synergistic value creation" and "next-generation service delivery models" trying to figure out which supplier was actually saying something substantive and which one had just hired a better proposal writer. You'd compile it all into a PowerPoint that took longer to format than it did to analyze, and you'd walk into that steering committee meeting running on coffee and prayer.

Instead, I did something I'd been experimenting with for a few months. I uploaded all three RFP responses into an AI tool. I asked it to extract and compare the key commercial terms across all three proposals in a standardized format. It did it in about ninety seconds. I asked it to identify the substantive differences in their service delivery models—not the marketing language, the actual operational commitments. Three minutes. I asked it to flag every instance where a supplier had been vague or noncommittal on a requirement we'd marked as critical, and to compare that against how specific the other two suppliers had been on the same requirement. Four minutes.

In under ten minutes, I had a comparative analysis that would have taken my team the better part of a week to produce manually. And it wasn't just faster—it was more thorough. The AI caught inconsistencies between suppliers' pricing appendices and their narrative descriptions that a human reviewer, fatigued after reading their third hundred-page document, would almost certainly have missed. It identified a liability limitation clause buried in one supplier's response that was significantly more restrictive than what the other two had proposed—the kind of thing that wouldn't matter until something went wrong, at which point it would matter enormously.

I sat there looking at my screen, and I had one of those moments that I suspect a lot of people in our profession have had recently, or are about to have. It wasn't fear. It wasn't excitement, exactly. It was more like the feeling you get when you've been driving the same route to work for twenty years and someone shows you a shortcut that cuts the trip in half, and your first thought isn't "great, I'll save time"—it's "how much time have I wasted?"

That's the moment this chapter is about. Not the technology itself—though we'll cover that, because you need to understand it well enough to use it and well enough to know when not to trust it. This chapter is about the mental shift. The recognition that the tools available to procurement professionals have fundamentally changed, that the change is accelerating, and that the procurement professionals who figure out how to work with these tools are going to have advantages over those who don't that are as significant as any competitive advantage I've seen in thirty years in this profession.

Let's get into it.

A note for the senior leaders before we go further:

The argument of Part Two is not that AI replaces procurement professionals. It's that AI replaces the manual work that has consumed most procurement professionals' calendars for the last thirty years (the data assembly, the document review, the report generation, the spreadsheet wrangling) and frees the talent you've already hired to do the work you actually hired them for. The strategic question for procurement leadership is not whether AI is real or whether it's coming. Both of those are settled. The question is whether your organization's procurement function is going to be one that captures the productivity multiplier this technology creates, or one that watches a competitor capture it. The chapters that follow are about how that capture happens. The talent you have can do this. They need permission, tools, and a leader who isn't waiting for the perfect AI strategy before starting.

THE PART WHERE I EXPLAIN AI WITHOUT MAKING YOUR EYES GLAZE OVER

Here's the thing about artificial intelligence as a topic of conversation: everybody has an opinion, almost nobody has a definition, and the gap between what people think AI does and what it actually does is wide enough to drive a procurement savings target through.

So let me give you the version that actually matters for your work. The procurement pro's version—what you need to understand to use these tools effectively, evaluate them critically, and avoid both the trap of over-reliance and the trap of dismissiveness.

At the broadest level, artificial intelligence is software that can perform tasks that traditionally required human intelligence. That's deliberately vague, and it should be, because the field encompasses everything from the algorithm that recommends your next Netflix show to the systems that are helping pharmaceutical companies identify drug candidates. When people in procurement talk about AI, they're usually talking about a much narrower set of capabilities, and it helps to know which ones actually matter for your work and why.

Machine Learning: The Pattern Recognition Engine

Machine learning is the foundation that most modern AI is built on. The core idea is deceptively simple: instead of programming a computer with explicit rules for how to do something, you feed it enormous amounts of data and let it figure out the patterns on its own.

A traditional program for categorizing procurement spend might have thousands of hand-coded rules: "if the vendor name contains 'Acme Office' and the GL code is 6500, categorize as Office Supplies." A machine learning model looks at millions of historical transactions that humans have already categorized and learns to recognize the patterns that determine which category a transaction belongs to—including patterns that no human programmer would have thought to code as a rule. Maybe transactions with certain combinations of dollar amounts, vendor types, and cost center codes follow patterns that are invisible to the human eye but crystal clear to the model.

Here's what this looks like in procurement practice.

I've seen organizations using machine learning for spend classification. Their model was trained on three years of historically classified transactions—roughly four million data points. After training, the model could classify new transactions with ninety-three percent accuracy. But the interesting part wasn't the overall accuracy rate. It was what the model found in the data that no human had noticed. The model identified that transactions from a specific set of cost centers, processed during the last week of each quarter, were being systematically miscategorized—coded to broad categories rather than specific ones. When the team investigated, it turned out that a quarterly budget reconciliation process was generating a surge of transactions that the AP team was rushing through without careful classification. The machine learning model spotted a pattern in four million records that no human analyst had ever seen—because no human analyst could look at four million records and spot a quarterly timing pattern in misclassification rates.

Machine learning in procurement shows up in spend classification, supplier risk scoring, demand forecasting, price prediction, anomaly detection, and invoice matching. Anywhere you have a large volume of data and a pattern recognition problem, machine learning is relevant. And procurement is, when you strip it down to its analytical core, a collection of pattern recognition problems with financial consequences.

Natural Language Processing: Making Machines Read

Natural language processing (NLP) is the branch of AI that deals with human language. Reading it, understanding it, extracting meaning from it, generating it. If machine learning is about finding patterns in numbers, NLP is about finding patterns in words.

This matters enormously for procurement because most procurement work involves processing text. Contracts. RFP responses. Supplier correspondence. Scope of work documents. Policy documents. Terms and conditions. Performance reports. Market research. Every one of these is a text-heavy document that someone has to read, understand, extract information from, and act on.

Consider contract analysis. A typical enterprise might have five thousand active contracts. Each contract contains dozens of key provisions—pricing terms, payment terms, liability limitations, termination clauses, auto-renewal provisions, service level definitions, escalation mechanisms. Traditionally, understanding what's actually in those contracts required a human being to read each one—or, more realistically, required the organization to simply not know what was in most of its contracts, because nobody had the time or resources to read them all.

NLP changes that equation. An AI system powered by NLP can read five thousand contracts, extract the key provisions from each one, normalize them into a comparable format, and present you with a portfolio-level view of your contractual commitments and exposures in a matter of hours. Not weeks. Not months. Hours. It can tell you how many of your contracts have auto-renewal clauses that will trigger in the next ninety days. It can tell you which contracts have liability caps below your organization's standard threshold. It can tell you which suppliers have the most favorable payment terms and which have the most restrictive. It can flag clauses that deviate from your standard template and quantify the financial impact of those deviations.

I used NLP-powered contract analysis last year to review a portfolio of forty-seven technology contracts ahead of a consolidation initiative. The manual approach would have required a paralegal and a procurement analyst working together for approximately three weeks. The AI produced the initial extraction and comparison in about four hours. The human review and validation—checking the AI's work, adding context, correcting the inevitable errors—took another day and a half. Total elapsed time: two days instead of three weeks. And the output was more comprehensive because the AI checked every clause in every contract, whereas a human team under time pressure inevitably prioritizes and skims.

Generative AI: The Capability That Changed Everything

Generative AI is the capability that has captured the world's attention, and for good reason. ChatGPT, Claude, Gemini, Copilot, and their rapidly multiplying cousins, can generate new text, code, images, and analysis in response to human prompts. They are, in essence, sophisticated pattern completion engines. They've been trained on vast amounts of human-generated text, and they've learned to predict what comes next in a sequence of words with a degree of sophistication that allows them to write coherent prose, summarize complex documents, analyze data, answer questions, and hold conversations that feel remarkably human.

Generative AI has the capability to turn every procurement professional with a laptop into someone who has a tireless, knowledgeable research assistant who has read everything, never sleeps, and can synthesize information at a speed that no human team can match. That's not hype. That's a description of what these tools can do right now, today, for the sourcing professional who knows how to use them.

Let me give you some concrete examples of what generative AI can do in a procurement context, because the range is broader than most procurement pros realize.

AI Capabilities:

- **Market research and supplier intelligence.** Ask a generative AI tool to analyze a supplier's most recent annual report and earnings call transcript, identifying their strategic priorities, margin pressures, and investment areas relevant to your negotiation. What used to require hours of reading and note-taking becomes a fifteen-minute analysis.
- **RFP development.** Give the AI your requirements document and ask it to draft an RFP that covers all functional requirements, includes appropriate evaluation criteria, and

follows your organization's standard format. The draft will need editing and refinement—it always does—but starting from a seventy percent complete draft rather than a blank page saves days.

- **Contract review.** Upload a supplier's contract redline and ask the AI to identify every deviation from your standard terms, categorize them by risk level, and suggest negotiation positions for the highest-risk deviations. The analysis won't replace your legal team's review, but it gives all those involved a head start that makes the review significantly more efficient.
- **Communication drafting.** Need to write a supplier notification about a change in requirements? A stakeholder update summarizing the status of a sourcing initiative? A response to a supplier's escalation? Generative AI can draft all of these, calibrated to the appropriate tone and level of detail, in minutes rather than the hour you'd normally spend staring at a blank email.
- **Data analysis and visualization.** Upload a spend data set and ask the AI to identify the top categories by spend, the suppliers with the highest year-over-year cost increases, and the business units with the most fragmented purchasing patterns. The analysis comes back in minutes, often with insights that would have taken your team a full day of spreadsheet work.

The pattern across all of these is the same. Generative AI compresses analytical, research, and communication work from hours or days into minutes. It doesn't eliminate human judgment—the AI's output always needs review, refinement, and the application of organizational context that only you can provide. But it changes the ratio of time spent on production versus thinking. The mechanical work compresses. The judgment work expands. That's the whole argument.

WHAT AI IS SPECTACULARLY BAD AT (AND WHY THAT MATTERS MORE)

I spend a lot of my time at Rutgers telling students something that is simultaneously the most important and least sexy thing about AI: understanding the limitations is more valuable than understanding the capabilities.

Capabilities are exciting. Limitations keep you out of trouble.

And in procurement (where the decisions you make have real financial, operational, and sometimes legal consequences) staying out of trouble matters.

Here's what current AI tools cannot reliably do, and what you should never fully outsource to them without human verification.

What AI CANNOT Reliably Do:

AI hallucinates. This is the technical term for when an AI generates information that sounds completely authoritative, is formatted beautifully, reads like it came from a senior analyst at Goldman Sachs, but it is **entirely fabricated**. It didn't find this information somewhere and get it wrong. It generated it from statistical patterns and presented it as fact because that's what its architecture is designed to do—complete patterns in ways that are plausible. I have personally watched AI tools generate supplier financial data that didn't exist, cite contract clauses that weren't in the document, and produce market benchmarks that were internally consistent, impressively specific, and completely made up.

Real World Example:

Let me tell you how this can go wrong in practice. A colleague of mine, a procurement director at a mid-sized company, used a generative AI tool to prepare market benchmarks for a software category negotiation. The AI produced a beautifully formatted comparison showing average market pricing for the category across three tiers of functionality. The numbers looked reasonable. They were specific. They came with what appeared to be source attributions. My colleague used them in the negotiation, citing "market research" that showed the supplier's pricing was fifteen percent above the median.

The supplier's account manager, who happened to be exceptionally well-prepared, asked where the benchmarks came from. My colleague cited the sources the AI had provided. The account manager pulled up those sources on their laptop, right there in the meeting. The sources existed, but they didn't contain the data the AI had attributed to them. The AI had fabricated the specific numbers and attached them to real-sounding citations. The negotiation didn't recover. The credibility damage took months to repair.

In a casual conversation, a hallucination is an embarrassment. In a procurement negotiation, it's a catastrophe.

If you walk into a supplier meeting and cite a market benchmark your AI tool fabricated, you have just destroyed the credibility we spent all of Chapter Two talking about as one of your most valuable professional assets. Always verify. Always cross-reference. Never cite an AI-generated fact in a high-stakes context without confirming it from a primary source.

AI doesn't understand your organization's context.

It doesn't know your company's risk tolerance, your CFO's priorities, your relationship history with a particular supplier, or the political dynamics that make a technically optimal recommendation practically impossible. It can analyze data brilliantly. It cannot exercise the organizational judgment that turns analysis into actionable recommendations. That judgment is yours. It will remain yours. And the moment you stop applying it because the AI's output looks polished enough to present without your own critical evaluation, you have made yourself less valuable, not more.

I learned this the hard way early in my AI experimentation. I asked a generative AI to recommend a sourcing strategy for a particular category based on the spend data and market information I'd provided. The AI came back with a perfectly logical recommendation to consolidate from five suppliers to two and run an aggressive competitive bid. On paper, it was textbook strategic sourcing. In reality, it would have been organizational suicide—because two of those five suppliers were politically protected relationships that reported to business unit presidents who sat on the executive committee, and any attempt to displace them without careful stakeholder management would have generated a firestorm that would have set the procurement function back two years.

The AI couldn't know that. It can only analyze what you give it. The organizational politics, the relationship history, the unwritten rules that govern how decisions actually get made in your specific company—that's your domain, and it's precisely where your value as a human professional is irreplaceable.

AI can be confidently wrong about things that matter.

When an AI tool gets a restaurant recommendation wrong, nobody gets hurt. When it gets a contract risk assessment wrong—when it tells you a liability clause is standard when it's actually unusually restrictive, or when it misreads an indemnification provision because the legal language was structured in an unusual way—the consequences are real and potentially severe. Current AI tools are remarkably good at processing standard patterns. They are less reliable with edge cases, unusual structures, and the kind of nuanced legal and commercial language that procurement professionals encounter regularly.

AI reflects the biases in its training data.

If the data an AI system was trained on reflects historical biases—and most data does—the AI will reproduce and sometimes amplify those biases. In procurement, this has practical implications for supplier diversity programs, for geographic sourcing decisions, and for any evaluation process where the AI's recommendations might systematically favor certain types of suppliers over others for reasons that have nothing to do with capability or value.

So, does this mean we shouldn't use it?

No. None of this means you shouldn't use AI. You should.

But you should use it the way a great procurement professional uses any tool: with clear understanding of what it does well, what it does poorly, where it needs supervision, and where your own judgment is the irreplaceable ingredient.

I've started calling this the "brilliant intern" model, and I find it resonates with sourcing professionals. Think of generative AI as the most brilliant, hardworking, well-read intern you've ever had—one who can research anything, summarize documents at superhuman speed, and draft analysis that's often impressively good. But they have no judgment. They have no organizational context. They will occasionally present made-up information with absolute confidence. And they need supervision—not micromanagement, but the kind of experienced oversight that catches the mistakes, adds the context, and turns raw analytical horsepower into actual business decisions.

You wouldn't let a brilliant intern present to your steering committee unsupervised. Don't let AI do it either.

PROMPT ENGINEERING: THE SKILL THEY DON'T TEACH IN PROCUREMENT SCHOOL (BUT SHOULD)

Here's something that surprises people when I say it, but it's true: the most important skill for getting value from AI in procurement is not technical. It's not data science. It's not coding. It's something that has come to be called prompt engineering—the art of asking AI the right questions, in the right way, with the right context, to get output that's actually useful rather than generically impressive but practically worthless.

And here's the thing every experienced procurement professional should hear clearly: you already have most of the skills that make someone a great prompt engineer. You just don't know it yet.

Think about what makes a great procurement professional effective when they're briefing an analyst or directing a consultant. They don't say "go research this category." That's a terrible brief and it will produce terrible work. They say something like: "I need you to analyze our facilities management spend across all North American locations for the last eighteen months. Break it down by service type—janitorial, HVAC, landscaping, security. Identify the top ten suppliers by spend, show me which locations are using which suppliers, and flag any locations where we're paying rates more than fifteen percent above the portfolio average. I need this by Thursday because I'm presenting to the VP of Operations on Friday."

That's a great brief: specific about the scope, defining the dimensions of analysis, setting the output format, providing context about why the work matters and when it's needed. And it will produce significantly better work from a human analyst than the vague "go research this" instruction.

Prompt engineering for AI is the same skill, applied to a different audience. The principles are identical: be specific about what you want, provide context about why you want it, define the output format, and give the AI enough background information to produce something useful rather than generic.

Let me give you some concrete examples, because this is one of those skills that's easier to demonstrate than to describe in the abstract.

Bad prompt: "Help me prepare for a supplier negotiation."

Good prompt: "I'm preparing for a contract renewal negotiation with a major cloud software vendor in the life sciences space. This is a seven-figure annual contract. The vendor has significant market dominance in their niche and we have high switching costs. Our current contract expires in ninety days. I need you to help me build a preparation dossier. Specifically: (1) What are the most common negotiation tactics that dominant SaaS vendors use during renewals, and how should I counter each one? (2) What leverage points typically exist for buyers in high-switching-cost technology renewals, even when competitive alternatives are limited? (3) What contract terms should I prioritize protecting or improving in a renewal where the vendor has significant pricing power? (4) Generate five creative deal structure options that could reduce our total cost while giving the vendor something they value—like a longer commitment, expanded scope, or a case study reference."

The second prompt will produce output that's orders of magnitude more useful than the first—not because the AI is smarter, but because the human asking the question gave it enough context and specificity to produce relevant analysis rather than a generic negotiation guide you could find in any business textbook.

Here are the principles of effective prompting I teach my Rutgers students and that transfer directly to procurement practice:

Be specific about context. Tell the AI what you're working on, what industry you're in, what the stakes are, and what constraints you're operating under. The more relevant context you provide, the more relevant the output will be. Don't make the AI guess what you need—tell it.

Define the output format. If you want a comparison table, say so. If you want bullet points, say so. If you want a narrative summary suitable for an executive audience, say so. If you want the analysis structured around questions, list the questions. AI tools are remarkably good at adapting their output to format specifications—but they can't read your mind about what format you want.

Assign a role. One of the most effective prompting techniques is to tell the AI what role to play. "Act as a senior procurement consultant with twenty years of experience in technology sourcing" will produce different—and usually more sophisticated—output than the same question asked without a role assignment. The role gives the AI a frame of reference for the level of expertise, vocabulary, and analytical depth you're expecting.

Iterate, don't one-shot. The best AI interactions are conversations, not single transactions. Start with a broad question. Review the output. Then drill deeper on the most promising areas. "That's helpful. Now take point number three and go deeper—give me specific contractual language I should be looking for and the typical negotiation dynamics around that clause." This iterative approach produces progressively more specific and useful analysis, much like a conversation with a human expert where each exchange builds on the last.

Provide examples. If you want the AI to produce something in a specific style or structure, show it an example. "Here's an executive summary I wrote for a previous sourcing initiative. I need you to produce a similar summary for this new initiative, matching the structure, tone, and level of detail." The AI will pattern-match to your example remarkably well, producing output that feels like your work rather than generic AI-generated content.

Challenge the output. Don't accept the first response uncritically. Ask the AI to identify weaknesses in its own analysis, to consider perspectives it might have missed, to play out the response—"what would a senior supplier account executive say in response to this negotiation strategy?" The AI can role-play both sides of a procurement interaction, which is valuable for stress-testing your preparation.

Prompt engineering is not a technical skill.
It's a communication skill.

And procurement professionals who spend their careers communicating complex requirements, directing analytical work, and extracting useful information from imperfect sources, are better positioned to develop effective prompts than almost anyone. The procurement professional who combines deep procurement expertise with strong prompting skills has a force multiplier that's transformative.

THE PROCUREMENT AI LANDSCAPE: WHAT'S REAL, WHAT'S HYPE, AND WHAT'S COMING

Now that we've established what AI can and can't do at a fundamental level, and how to get the most out of it through effective prompting, let's talk about how it's actually showing up in procurement. The landscape is moving fast, and it helps to have a map.

There are essentially three layers of AI capability that are relevant to procurement professionals today.

Layer One: AI Embedded in Your Existing Tools

If you're using any modern procurement platform (a source-to-pay suite like Coupa, SAP Ariba, or Jaggaer, a contract lifecycle management system, a spend analytics tool, or a reasonably current ERP) there's a good chance that AI and machine learning capabilities are already built into the software you're using every day. Automated spend classification. Intelligent invoice matching. Predictive supplier risk scores. Recommended sourcing strategies based on historical patterns. Community-sourced benchmarking data that uses AI to normalize and compare across thousands of organizations.

These capabilities have been shipping in enterprise procurement tools for several years and getting meaningfully better with each release. Coupa's AI analyzes community spend data across thousands of companies to provide pricing benchmarks and risk scores that individual organizations couldn't generate on their own. SAP Ariba has embedded generative AI through its Joule copilot to help with category management recommendations and natural language search across procurement data. Zycus has built a generative AI suite (Merlin) that integrates with Microsoft Teams to provide conversational AI assistance for procurement tasks.

The important point: most procurement professionals are already using AI whether they realize it or not. If your spend management platform automatically classifies transactions, you're using machine learning. If your contract system flags upcoming renewals based on pattern analysis, you're using AI. The question isn't whether you're using AI —it's whether you're using it deliberately and strategically, or whether it's just running quietly in the background while you continue doing things the old way.

Layer Two: General-Purpose AI Applied to Procurement

This is where ChatGPT, Claude, Gemini, Copilot, and their peers come in. These are not procurement-specific tools. They're general-purpose AI assistants that happen to be remarkably useful for procurement work when wielded by someone who knows what they're doing.

Market research. Supplier analysis. Contract review. RFP development and evaluation. Negotiation preparation. Communication drafting. Data analysis. The range of procurement tasks where a skilled procurement professional can get meaningful value from a general-purpose AI tool is broad, and this is the layer where the democratization of capability is most pronounced. You don't need a six-figure software license or a data science team. You need a laptop, an AI subscription that costs less than your monthly coffee budget, and the knowledge to use it effectively.

The logistics negotiation story from Chapter Two—where I used a generative AI tool to synthesize earnings call transcripts, benchmark pricing, and identify that the supplier had lost two major accounts in our region—was a layer two application. I wasn't using a procurement-specific tool. I was using a general-purpose AI assistant and applying procurement expertise to direct it. The expertise was the differentiator, not the technology.

This is the layer I encourage every procurement professional to start with, regardless of what technology their organization has formally adopted. It's accessible, affordable, and the skills you develop here, particularly the prompt engineering skills we just discussed, transfer directly to every other layer of the AI landscape.

Where to Start: The Monday Morning Entry Point

If you've never used a generative AI tool for procurement work, here's what I'd suggest you do this week. Pick one tool—ChatGPT, Claude, Gemini, CoPilot... whichever your organization permits—and pick one workflow you do regularly that involves reading or comparing documents. RFP responses. A supplier's annual report. A handful of contracts. Last quarter's spend report. It doesn't matter which one. Choose the one you hate doing the most. (It might make it fun.)

Upload it. Ask the AI three questions you would normally answer manually. Compare what it produces to what you would have produced. Don't grade the AI on a pass/fail basis—grade it on time saved, on what it caught that you would have missed, and on what it missed that you would have caught. The point is not that the AI is perfect. It isn't. The point is that you're learning the texture of the tool—where it helps, where it doesn't, what kind of prompt produces what kind of output.

Do this with one workflow. Then another. Within a month of disciplined experimentation, you'll have a working sense of where AI delivers value in your specific work and where it doesn't. That working sense is the foundation everything else in this book builds on. The best buyers who develop it early have an advantage that compounds over time. The ones who keep waiting for the perfect moment will start the same experimentation later, with the same learning curve, just behind.

Layer Three: Purpose-Built Procurement AI Solutions

This is the rapidly emerging category of tools specifically designed to apply AI to procurement use cases. It's where the most exciting and transformative applications are being built. Let me walk you through what's out there, because the landscape has evolved significantly even in the past year.

AI-powered spend analytics platforms like Suplari and the spend intelligence modules in major S2P suites can automatically classify millions of transactions, normalize suppliers, detect anomalies. These tools go beyond what a general-purpose AI can do with spend data because they're trained specifically on procurement transaction patterns and have been refined across hundreds of implementations.

Supply chain risk monitoring platforms like Resilinc use AI to continuously scan global news sources, financial filings, social media, weather data, and regulatory databases to provide real-time early warning of supply chain disruptions. Their EventWatch AI capability monitors events worldwide and alerts procurement teams to potential impacts on their specific supply chains—a capability that was science fiction a decade ago and is now operational at hundreds of major companies.

Contract intelligence platforms are using NLP to transform contract management from a filing cabinet exercise into a source of strategic insight. These tools can ingest thousands of contracts, extract key terms, identify risk patterns, benchmark clauses against market standards, and flag upcoming renewals and expirations. The analysis that used to require a team of paralegals working for weeks can now be produced in hours.

Intake and orchestration platforms like Omnea and Tonkean are using AI to transform how procurement requests enter the system. Instead of manual forms and email-based routing, these tools use natural language processing to interpret purchase requests, automatically classify them, apply policy rules, route them to the appropriate workflow, and even generate RFPs for straightforward requirements. Tonkean's ProcurementGPT capability can auto-generate RFPs, compare vendor options, and create compliant purchase orders for tail spend—the kind of low-value, high-volume purchasing that has traditionally consumed a disproportionate amount of procurement team bandwidth.

Autonomous sourcing and negotiation tools represent the cutting edge. Platforms like Procure AI have built AI agents that can execute tactical sourcing events and negotiations autonomously for routine, well-defined categories—sending RFQs, evaluating responses, conducting automated negotiations based on pre-set parameters, and generating awards. Procure AI reports that customer outcomes include thirty-seven percent shorter order processing times, forty-seven percent faster award decisions, and average savings of nearly five percent on tail spend negotiations. The human procurement professional sets the strategy, defines the parameters, and reviews the outcomes—but the execution is handled by AI agents that can operate at a speed and scale that no human team could match.

Procurement pro-built platforms are another emerging dimension worth watching. *Full disclosure: ProcureOptima, which I'll mention briefly in this section, is the platform I built.* Tools like ProcureOptima are designed specifically around the workflows and productivity challenges that procurement professionals actually face, rather than the theoretical use cases that technology vendors imagine. These platforms prioritize sourcing productivity and strategic decision support, reflecting the insight that the people who best understand procurement's technology needs are the people who do procurement work every day. As more procurement teams with dual expertise in procurement and technology enter the market, I expect this category to grow rapidly.

The layer three landscape is evolving at a pace that makes any specific product recommendation likely to be outdated within a year. What won't be outdated is the framework for evaluating these tools: Does the platform solve a problem that's genuinely painful for your organization? Does it integrate with your existing technology stack, or does it create another data silo? Is the AI transparent enough that you can understand and validate its recommendations? Does it augment your team's capability, or does it require you to restructure your processes around the tool's limitations? And—perhaps most importantly...

Does the vendor understand procurement deeply enough to build something that works in the real world, or are they technology people who think they understand procurement because they read about it?

Those evaluation questions matter more than any feature comparison or vendor demo.

THE FEAR QUESTION: "IS AI GOING TO TAKE MY JOB?"

Let's address this directly, because I know it's on people's minds, and pretending it isn't would be dishonest.

Some procurement jobs are going to be eliminated by AI. I'm not going to sugarcoat that. The roles most vulnerable are the ones primarily transactional and repetitive—purchase order processing, basic invoice matching, routine catalog buying, standard report generation, and the manual data cleansing and classification work that has consumed so much of procurement's bandwidth for so long. These are tasks that AI and automation can perform faster, more accurately, and at significantly lower cost than humans. For organizations using expensive procurement professionals to perform these tasks, the economic case for automation is overwhelming.

But here's the distinction: those are tasks, not the profession. And the profession (the strategic, judgment-intensive, relationship-driven discipline we've spent the first three chapters describing) is not being automated. It's being empowered.

Think about what happened when spreadsheet software replaced manual accounting. Did it eliminate the accounting profession? No. It eliminated the manual calculation tasks that accountants used to spend most of their time on, and it freed them to focus on the analytical, advisory, and strategic work that was always more valuable but that they never had time for because they were too busy adding up columns of numbers by hand. The accountants who embraced spreadsheets became more valuable. The ones who insisted that real accounting meant doing the math on paper became irrelevant. The same dynamic is playing out in procurement right now, with AI as the catalyst.

The procurement professionals embracing AI are discovering that they can do in hours what used to take weeks—and the time they're freeing up allows them to focus on the high-value activities that have always been where procurement creates the most impact: deeper supplier relationships, more creative negotiations, better risk management, and strategic advisory partnerships with business stakeholders who are starting to see procurement not as a processing function but as a source of real competitive insight.

Here's a concrete example.

A category manager I know was spending roughly sixty percent of her time on data collection, spend analysis, and report preparation—the mechanics of building the analytical foundation for her sourcing strategies. The remaining forty percent was spent on the actual strategy development, stakeholder engagement, and supplier negotiations that created the most value. After adopting AI tools for her analytical workflow, the ratio flipped. Data collection and analysis now takes about fifteen percent of her time. The remaining eighty-five percent is spent on strategy, relationships, and execution. She's covering more categories, delivering better outcomes, and, her words, "finally doing the job I was hired to do instead of the data entry I was actually doing."

If your value as a procurement professional is primarily in processing transactions and producing reports, you should be concerned—not about AI specifically, but about the trajectory of the profession, which has been moving away from that model for decades. AI is accelerating a transition that was already underway.

If your value is in judgment, relationships, strategy, creativity, and the ability to navigate complex business situations with incomplete information (in other words, if your value is in the things we've been talking about in the beginning of this book) then AI isn't a threat. It's a superpower. The sooner you start developing the skills to wield it, the more formidable you're going to become.

THE DATA QUESTION: PROTECTING WHAT MATTERS

Before we wrap, a word about data security and confidentiality, because every procurement professional needs to think carefully about this when using AI tools.

Procurement professionals handle sensitive information every day. Supplier pricing. Contract terms. Internal cost targets. Strategic plans. Competitive intelligence. Some of this information is subject to NDAs. Some is subject to regulatory requirements. All of it has competitive value that would be damaged if exposed to the wrong parties.

When you upload a contract to an AI tool, paste supplier pricing data into a chatbot for benchmarking, or share strategic documents with an AI assistant for summarization, you need to understand where that data goes, how it's stored, whether it's used to train the AI model (which could potentially expose it to other users), and whether your organization's data governance policies permit this use.

The practical guidance: know your organization's AI usage policy, and if one doesn't exist, advocate for creating one. Understand the data handling practices of any AI tool you use—does the provider use your inputs to train their models? Most enterprise-grade AI offerings now offer options where your data is not used for training, but verify rather than assume. Be thoughtful about what data you share. Use anonymized or redacted data when specific details aren't necessary for the analysis. Don't paste a supplier's confidential cost breakdown into a public AI chatbot. Apply the same judgment about data sensitivity you would apply to sharing information with an external consultant—because that's essentially what you're doing.

Treat AI data governance as an extension of your existing information security practices. Same risk discipline, applied to a new dimension.

THE MINDSET SHIFT: FROM SKEPTIC TO SOURCING PROFESSIONAL

I want to close this chapter the same way I closed the first three—with a mindset observation, because technology adoption is ultimately a human behavior challenge, not a technical one.

The procurement professionals I know who are getting the most value from AI right now share a common trait, and it's not that they're technology enthusiasts. Some of them are. Some of them are self-described technophobes who resisted the move from paper to email twenty years ago. What they share is a willingness to experiment without needing certainty. They don't wait until they fully understand the technology before they start using it, until their organization has an official AI strategy, or until someone builds them a custom procurement AI platform with their exact taxonomy and their exact workflows. They start. They open an AI tool, throw a procurement problem at it, evaluate the output, learn what works and what doesn't, and iterate. Every day, they get a little better at asking questions, a little faster at recognizing when output is good and when it's nonsense, a little more creative about finding new applications.

They treat AI the way the best procurement professionals have always treated any new capability: as something to be tested, evaluated, and either adopted or discarded based on whether it delivers real value in real situations. Not based on vendor demos. Not based on conference keynotes. Not based on articles in procurement magazines that use the word "transformation" fourteen times without ever explaining what actually transformed. Based on whether it makes their work better.

There's a parallel here to the negotiation mindset we talked about in Chapter Two. The best negotiators are curious about the other party. The best AI adopters are curious about the technology. Not because they're tech enthusiasts—because they're professionals who recognize that understanding a powerful new tool is part of their job, just like understanding a new market or a new supplier is part of their job. Curiosity is the common thread. It's what makes sourcing professionals adaptive, and adaptive buyers are the ones who thrive when the landscape shifts.

That's the spirit I'd like you to bring forward. Not as a true believer who thinks AI will solve all of procurement's problems. Not as a skeptic who thinks it's all hype and that the fundamentals are all that matter. Read it as a professional who has mastered the fundamentals, who understands the tools are changing, and who is ready to learn how to use the new tools with the same rigor and judgment they bring to everything else.

All my life I've been told that AI was the bad guy. From HAL 9000 wanting to kill everybody on the Discovery One, to Skynet deciding humanity needed to be terminated, to Ultron spending five minutes on the internet and concluding we all had to go. (I can't say I blame him—I know what some of you are into.) Hollywood has spent sixty years telling us that artificial intelligence is the existential threat we should all be terrified of.

Turns out, the reality is a lot more mundane and a lot more useful. AI in procurement isn't Skynet. It's not Rosie from the Jetsons either—the seamless, do-everything robot that integrates perfectly into your life without any effort on your part. The reality is somewhere in between: a powerful set of tools that require real skill to use effectively, that have real limitations you need to understand, and that are creating opportunities for the procurement professionals who invest in learning them that the Big Rons and the 1994 conference room negotiators could never have imagined.

The machines are in the room now. They're not going anywhere. The only question is whether you're going to learn to work with them or watch someone else do it first.

Let's go learn how.

CHAPTER 4
SPEND ANALYSIS SUPERCHARGED

SEEING WHAT YOU'VE NEVER SEEN BEFORE

I once inherited a spend analysis that told me my organization was spending $4.2 million annually with a supplier I'd never heard of. The name looked vaguely like a facilities management company, but nobody in procurement had a contract with them, nobody in accounts payable recognized the vendor code, and when I asked the finance team about it, they shrugged and said it had been coded that way for years.

It took my team three weeks to figure out what was going on. The $4.2 million wasn't one supplier. It was sixty-seven different suppliers whose invoices had been miscoded into the same vendor master record by a well-meaning accounts payable clerk who had been with the company since the Reagan administration and had developed her own personal taxonomy for categorizing vendors because nobody had ever given her one. She had decided, at some point in the distant past, that anything she couldn't immediately classify went into a catch-all account she'd created and labeled with a name that was, as near as I could determine, a misspelling of her cat's name.

Sixty-seven suppliers. $4.2 million. Coded under what was essentially a filing system named after a cat called Mr. Whiskers.

If you've worked in procurement at any organization larger than a lemonade stand, you have your own version of this story. Maybe yours involves the supplier who appeared in your spend data under fourteen different name variations—"IBM," "I.B.M.," "International Business Machines," "IBM Corp," "IBM Corporation," "IBM Global Services," and eight more creative spellings that somebody in your AP department invented on a Tuesday afternoon when they couldn't be bothered to look up the master record. Maybe yours is the discovery that thirty percent of your indirect spend was flowing through purchasing cards with no category coding at all, which meant that as far as your spend analytics were concerned, it simply didn't exist. Maybe yours is the quarterly spend report that took six people four weeks to produce and was already two months out of date by the time anyone saw it.

This is the state of spend analysis at most organizations. Procurement spend data is messy, fragmented, inconsistently coded, spread across multiple systems, and generated by thousands of people who have no particular incentive to code their purchases in a way that makes a procurement analyst's life easier. Turning that raw data into something you can actually make strategic decisions from has been, for the entire history of the profession, an exercise in grinding, tedious, manual effort that never quite produces the visibility you actually need.

Until now.

Spend analysis, the ability to see and understand what your organization is actually spending, is one of the areas where AI is delivering the most immediate and dramatic value in procurement in ways that are fundamentally changing what's possible for procurement teams who know how to use the tools.

Let me show you what I mean.

If you run a procurement organization, the strategic case for this chapter compresses to one idea. Spend visibility is the foundation under which every other procurement value-creating activity sits—sourcing strategy, supplier consolidation, risk management, contract compliance, category development. Without it, everything else is approximation. The historical reality is that most enterprises have never had real spend visibility, because the manual work required to produce it was either prohibitively expensive or impossible at scale.

AI changes the economics. The question is not whether to invest in AI-powered spend analytics. It's whether your organization is going to be one that operates from real spend visibility going forward, or one that continues making strategic decisions about supplier consolidation, sourcing prioritization, and category strategy based on data that is months old, partially classified, and structurally incomplete. The chapter that follows is about how the first option becomes possible.

A note on what follows.

The time-savings numbers, savings examples, and outcome stories in the chapters that follow come from my own work with procurement organizations and from what I've watched happen across multiple engagements over the past few years. They are illustrative of what skilled application of these tools can deliver. They are not findings from controlled studies. Where I cite specific research or name a source, the source is public and you can check the numbers. Everything else is what a thirty-year operator has seen in the field, sanity-checked against what other operators I trust are seeing in theirs.

Treat the examples as proofs of concept. Your mileage will vary—based on your tools, your data, your team, and the situations you apply them to. The point isn't to promise you specific numbers. The point is to show you the shape of what's possible and give you the workflows to find out what the numbers actually are in your context.

If you've seen Moneyball, you've seen the spend analysis problem. The Oakland A's had the same statistics every other team had—sitting in the same record books, available to anyone who wanted to look. What Billy Beane and Paul DePodesta did wasn't acquire better data. They asked better questions of the data that was already there. The hidden value was sitting on the page in front of every general manager in the league. No proprietary feed, no exclusive access, just data nobody else had bothered to interrogate. Spend analysis is the same problem. The data is in your ERP, your accounts payable system, your purchasing card statements, your supplier master records. It has been there for years. The question is whether anyone has the time, the tools, and the patience to ask it the right questions.

WHY SPEND VISIBILITY IS THE FOUNDATION OF EVERYTHING

Before we get into what AI changes, I want to make sure we're aligned on why this matters, because spend analysis, of all the procurement disciplines, is the one most likely to make your eyes glaze over. It doesn't have the drama of a high-stakes negotiation. It doesn't have the strategic gravitas of a supplier relationship transformation. It doesn't have the intellectual satisfaction of a creative contract restructuring. Spend analysis is, at its core, data work. It's categorization and normalization and anomaly detection and reporting. It's the vegetables on the procurement plate —essential for your health, but nobody's writing poetry about it.

And yet every impressive procurement outcome I've ever delivered or witnessed, every creative sourcing strategy, every breakthrough negotiation, every risk averted just in time, started with spend visibility. Every one. Because you cannot develop a sourcing strategy for a category if you don't know how much you're spending in that category, with which suppliers, across which business units, and under what contractual terms. You cannot consolidate spend to create negotiating leverage if you don't know the spend is fragmented in the first place. You cannot identify maverick spending (purchases that flow outside of your preferred channels and negotiated agreements) if you can't see where the money is actually going. You cannot measure the impact of your procurement initiatives if you don't have a reliable baseline to measure against.

Spend visibility is to procurement what medical imaging is to medicine. Before imaging technology, doctors were diagnosing based on symptoms, patient history, and physical examination—useful tools, but fundamentally limited. They were making life-and-death decisions without being able to see inside the patient. Imaging changed that. It didn't make the doctor's judgment less important—it made it more effective by giving them information they'd never had before.

That's what great spend analysis does for procurement. And that's what AI is transforming.

THE OLD WORLD: WHY TRADITIONAL SPEND ANALYSIS WAS A NIGHTMARE EVERYONE TOLERATED

You can't appreciate the transformation without understanding just how painful the status quo actually is, so let me take you inside the process, in case you've been fortunate enough never to run a traditional spend analysis from scratch.

Traditional spend analysis follows a process that hasn't fundamentally changed in twenty years. You start by extracting data—pulling transaction records from your ERP, your procurement systems, your purchasing card platforms, and whatever other systems your organization uses to buy things. This sounds simple, but it's not. Most organizations of any size have multiple ERP instances, multiple procurement platforms, and a disturbing number of transactions that flow through systems that procurement doesn't control or even have visibility into.

At one Fortune 500 company I worked with, just getting the raw transaction records into a single file took two weeks because the data lived in seven different systems across four different countries, each with its own coding conventions, currency denominations, and data formats. The European subsidiary coded suppliers by their VAT number. The North American operations used an internally generated vendor code. The Asian operations used a third system entirely. Before we could even begin analyzing the spend, we had to build a translation layer that allowed these systems to talk to each other. That translation layer, incidentally, was a massive Excel spreadsheet maintained by one person who was, I was gently informed, approaching retirement and had no documented process for how any of it worked.

This is not unusual. This is Tuesday.

Once you've extracted the data, you clean it. This is where the real suffering begins. Supplier names need to be normalized.

- All fourteen variations of IBM need to be recognized as the same entity.
- Transactions need to be categorized—that office furniture purchase that was coded to "miscellaneous" needs to be moved to the right spend category.
- Duplicate transactions need to be identified and removed.
- Currency conversions need to be applied.

- Business unit and cost center mappings need to be verified.
- Exceptions and anomalies need to be investigated.

In a manual environment, this cleansing and classification process is mind-numbing work. You're looking at rows of transaction data (hundreds of thousands of them, sometimes millions) and making judgment calls about which category each one belongs in. Is that $15,000 payment to "Acme Professional" a consulting engagement, a staffing agency placement, or a training services purchase? The transaction description says "services rendered per agreement." Tremendously helpful. Now multiply that ambiguity by half a million transactions, add the fact that your team has been staring at spreadsheets for three weeks and their judgment is starting to deteriorate along with their will to live, and you understand why accuracy rates in manual spend classification hover around seventy-five to eighty-five percent on a good day.

At one organization I worked with, the annual spend analysis took a team of four people approximately ten weeks to produce. By the time it was complete, the data was already three to four months old, the spend patterns had shifted, and the analysis was more of a historical record than a decision-making tool. It was useful—any visibility is better than none—but it was a snapshot from the recent past, not a picture of what was happening now. It was like trying to drive by looking in the rearview mirror. You could see where you'd been, but it told you almost nothing about what was directly in front of you.

And then came the presentation. The moment where you'd walk into a meeting with your Chief Procurement Officer or your CFO and present the results. And inevitably, someone would challenge a number. "That can't be right. There's no way we're spending $8 million on temporary staffing. Where's that coming from?" And you'd have to trace it back through the data, and sometimes the number was right and the stakeholder was wrong, and sometimes the number was wrong because a classification error had lumped two categories together, and you'd spend the next week running corrections and reissuing the analysis and watching whatever credibility you'd built with that stakeholder evaporate in real time.

I don't miss that world. Nobody should.

THE DATA QUALITY PROBLEM: WHY IT'S WORSE THAN YOU THINK

Before we jump to the AI solution, I want to spend a moment on something that doesn't get enough honest attention in the procurement technology conversation: data quality. The dirty secret of spend analysis, AI-powered or otherwise, is that the quality of your output is fundamentally constrained by the quality of your input. And at most organizations, the quality of procurement transaction data is, to use a technical term, terrible.

Here's why. Procurement transaction data is generated by a vast number of people across the organization, most of whom have no training in data management, no incentive to code their purchases accurately, and no idea that their sloppy coding is going to cause a procurement analyst to question their life choices six months later.

Consider:

- The sales rep who submits an expense report for a client dinner doesn't think about whether they're using the right expense category.
- The IT manager who buys a software license through a purchasing card doesn't think about whether the transaction description is detailed enough for someone else to classify it correctly.
- The administrative assistant who processes an invoice doesn't think about whether they're entering the supplier name in the same format that the previous three invoices used.

Multiply these small acts of data negligence by the thousands of transactions that flow through a large organization every day, and you get a data set that is, by the time it reaches the procurement analyst's desk, a masterwork of inconsistency. Supplier names are spelled differently across transactions. Category codes are applied inconsistently or not at all. Transaction descriptions range from precisely detailed to a single word that could mean anything. And a meaningful percentage of transactions (often twenty to thirty percent) are coded to catch-all categories like "miscellaneous" or "other."

Or, in the case of one particularly creative organization I encountered, a category called:

"ask Dave."

I have spent more hours of my professional life than I

care to count dealing with the downstream consequences of upstream data quality failures. It is, without exaggeration, the single biggest practical challenge in procurement analytics. And it is the context in which AI's contribution to spend analysis needs to be understood—not as a magic wand that makes bad data good, but as a set of capabilities that can dramatically improve the speed and accuracy of working with imperfect data while still requiring human judgment about what the data means.

I know what some of you are thinking right now. Spend analysis is the part of procurement that nobody goes home excited about. The dinner-party version of this chapter would not get you invited back. Push through anyway. The chapters that look more exciting depend on this one being done right, and the operators who treat spend visibility as the unglamorous foundation are the ones who deliver the glamorous results.

THE AI TRANSFORMATION: WHAT CHANGES AND HOW

Now let's get into the substance. AI brings a set of capabilities to spend analysis that, taken together, transform it from a periodic, painful, backward-looking exercise into a continuous, real-time strategic capability. Let me walk you through each one in detail.

Automated Spend Classification

This is the foundational capability, and it's the one that delivers the most immediate return on investment.

Machine learning models can be trained to classify procurement transactions into spend categories with accuracy rates that consistently exceed what human analysts achieve—typically ninety to ninety-five percent accuracy, compared to the seventy-five to eighty-five percent that's typical of manual classification.

Here's how it works in practice. You feed the model your historical transaction data—the transactions that have already been classified by human analysts, however imperfectly. The model learns the patterns: which combinations of supplier name, description, GL code, cost center, and other attributes correspond to which spend categories. Once trained, it can classify new transactions automatically, in real time, as they enter the system. No more quarterly batch processing. No more ten-week projects. Transactions get classified as they happen, as fast as they flow through your systems.

But the real power isn't just speed and accuracy—it's coverage. In most organizations, procurement spend analysis covers somewhere between fifty and seventy percent of total organizational spending. The rest is either in categories that procurement doesn't traditionally manage, in systems that procurement doesn't have access to, or in transactions that are too small, too fragmented, or too poorly coded to classify. AI-powered classification can push that coverage to ninety percent or higher, because the model doesn't get tired, doesn't skip the small transactions, doesn't decide that the $200 purchase isn't worth classifying, and can process data from any source you can connect it to.

The thirty to forty percent of spend that was previously

invisible is where some of the biggest sourcing opportunities hide. I have seen organizations discover millions of dollars in spend with suppliers they didn't know they were using, in categories they didn't realize were that large, flowing through channels they didn't know existed. At one company, an AI-powered spend analysis revealed that the organization was spending over $3 million annually on translation services—a category that didn't even exist in their spend taxonomy because the spend had always been coded to "professional services," where it was invisible among the much larger consulting and legal spend. Once visible, it took about six weeks to run a competitive sourcing event that reduced the cost by thirty-five percent and consolidated from eleven suppliers to two. That savings had been sitting there for years. Nobody could act on it because nobody could see it.

You cannot negotiate what you cannot see. And AI-powered classification lets you see things that were previously invisible.

The models also learn and improve over time. When a human analyst corrects a misclassification, that correction feeds back into the model, improving its accuracy for similar transactions in the future. This creates a virtuous cycle where the model gets progressively better as it processes more data and receives more human feedback. An organization that starts at ninety percent accuracy after initial deployment might be at ninety-five or ninety-six percent after a year of use as the model learns the patterns and exceptions in that organization's data.

One thing I want to be honest about: the initial training period matters. A model trained on a large data set of well-classified historical transactions will perform significantly better than one trained on a small, poorly classified data set. Organizations with a history of rigorous spend classification have a head start. Organizations whose historical data is a mess (and let's be honest, that's most organizations) will need to invest more effort in the initial data preparation and will see lower accuracy rates initially. The AI accelerates the journey to good spend visibility, but it doesn't eliminate the need for the journey itself.

Supplier Normalization

Remember the fourteen versions of IBM? AI handles this with a capability called entity resolution. It recognizes that "IBM," "I.B.M. Corp," "International Business Machines Corporation," and "IBM Global Services - Chicago" are all the same parent entity.

Modern AI-powered supplier normalization doesn't just match on name strings. It cross-references against external databases of corporate entities, considers addresses and tax identification numbers, analyzes transaction patterns, and builds a unified supplier hierarchy that shows you not just your total spend with a company, but your spend with each subsidiary, division, and business unit—and how those relate to each other.

The practical impact is significant, and I want to give you a detailed example because this is one of those areas where the value of AI often surprises procurement professionals who've never seen it in action.

A detailed example:

A consulting client of mine discovered they had relationships with roughly three thousand suppliers. This number came from their vendor master—the official list of suppliers in their ERP system. Based on this number, they believed their supply base was already relatively consolidated and that there wasn't significant opportunity for further rationalization.

After AI-powered supplier normalization, the actual number of unique supplier entities was closer to eleven hundred. Nearly two-thirds of the entries in their vendor master were duplicates, subsidiaries, or different business units of suppliers they were already working with. And here's where it gets interesting: the normalization didn't just reduce the number of suppliers on the list. It revealed patterns of fragmentation that had been completely invisible in the un-normalized data.

Three separate business units had been buying essentially the same professional services from three different subsidiaries of the same consulting firm, each negotiating independently, each paying different rates, and none of them aware of the others' contracts. One division was paying $275 per hour for the same capability that another division was getting for $195 per hour from a different subsidiary of the same parent company. The consolidated view made it immediately obvious that a single enterprise agreement would deliver better terms—which it did. The rate differential alone, applied across the combined volume, was worth over $400,000 annually.

But the benefits went beyond pricing. The consolidated view also revealed that this consulting firm had the broadest exposure to their organization of any supplier—they were working across multiple business functions, with access to sensitive strategic and operational information in each. From a risk management perspective, that concentration of access was something the CISO needed to know about, and it was completely invisible in the fragmented vendor master. The supplier normalization exercise that started as a spend analytics initiative ended up triggering a vendor access review that identified and remediated several information security gaps.

That's the cascading value of visibility. You start by cleaning up supplier names, and you end up discovering pricing discrepancies, consolidation opportunities, and risk concentrations that were all hiding in plain sight.

Anomaly Detection

This is where AI starts to feel less like an efficiency tool and more like having a financial detective on your team. Machine learning models are exceptionally good at identifying patterns in data, which means they're also exceptionally good at identifying when something breaks the pattern.

In spend analysis, anomaly detection flags transactions that don't fit the normal pattern for their category, supplier, business unit, or time period.

Anomaly detection can flag:

- An invoice from a regular supplier that's forty percent higher than their typical invoice.

- A sudden spike in spending in a category that's normally flat.
- A new supplier that starts receiving large payments without going through the standard onboarding process.
- A pattern of transactions just below the approval threshold that might indicate someone is deliberately splitting purchases to avoid oversight.
- A supplier whose invoice frequency has doubled even though the contract terms haven't changed.

Some of these anomalies have innocent explanations. A spike in facilities spending might reflect a planned office renovation. A new supplier might be legitimate but onboarded through an expedited process that skipped some documentation steps. The point isn't that every anomaly is a problem. The point is that AI can scan millions of transactions and surface the ones that warrant human investigation—something that manual review processes simply cannot do at scale. A human reviewer looking at invoices can process maybe a few hundred per day with reasonable attention. An AI model can scan millions of transactions in minutes and flag the fifty that look unusual. Those are very different capabilities, and the difference matters for organizations that process tens of thousands of invoices monthly.

I know a CPO who credits AI-powered anomaly detection with catching a billing error from a technology vendor that had been overcharging by approximately $180,000 per quarter for nearly two years. The overcharge wasn't dramatic enough on any single invoice to trigger a manual review. Each invoice was only about eight percent higher than it should have been, well within the range that a busy AP team would approve without question. But the AI model identified that this vendor's invoices had been trending upward at a steady, almost imperceptible creep that had been going on for seven consecutive quarters. No human reviewer had spotted the pattern across the hundreds of invoices they were processing monthly, because no human reviewer was comparing this quarter's average invoice amount to last quarter's and the quarter before that and the quarter before that. The AI was. And that seven-quarter trend, once surfaced, was unmistakable.

The recovery was significant—nearly $1.4 million in overcharges, plus a correction to the ongoing billing that saved another $720,000 annually going forward. That's $2.1 million in value from a single anomaly detection flag. The CPO told me, with the kind of understatement that procurement professionals specialize in, that the AI tool "paid for itself that week."

And then there's the fraud detection dimension. I want to be careful here, because accusing people of fraud based on data patterns is dangerous territory, and I am not suggesting that AI should be making those determinations. But AI-powered anomaly detection can identify patterns that would be virtually impossible to detect through manual review, and warrant investigation.

Fraud markers AI can see easily:

- A supplier that receives a disproportionate number of round-dollar invoices just below the approval threshold.

- A purchasing card user whose transaction patterns shift dramatically on certain days of the week.
- A supplier that was created in the vendor master the same week they received their first payment, with no evidence of a sourcing process or contract.

These patterns don't prove anything. But they're the kind of signals that procurement and internal audit should be investigating, and without AI-powered detection, most of them go unnoticed.

Real-Time Visibility

Perhaps the most transformative shift is the move from periodic, backward-looking spend analysis to continuous, real-time spend visibility. When classification and normalization are automated, and when your spend analysis platform is connected to your transaction systems via integration or data pipeline, you don't need to wait for the quarterly spend report. You can see what's happening now.

This changes the nature of procurement conversations with the business in ways that are difficult to overstate. Let me give you an example.

In the old model, I would walk into a quarterly business review with a divisional VP carrying a spend report that showed what their division had spent in the previous quarter. The data was already two to three months old. The VP might challenge a number—"we definitely didn't spend that much on temporary staffing"—and I'd have to go back, investigate, and come back the following week with an explanation. The conversation was inherently backward-looking and often defensive. I was presenting history, and the stakeholder was either accepting it or disputing it. Neither of us was in a position to do anything about what had already happened.

In the new model, with real-time spend visibility, I walk into that same conversation with a dashboard that reflects this week's transactions. I can show the VP not just what they spent last quarter, but what they're spending right now, this month, this week. I can show them that their temporary staffing spend is trending twenty percent above the run rate we discussed at the beginning of the year, and that if the current trajectory continues, they'll be $600,000 over budget by year end. I can show them which hiring managers are driving the increase, which staffing agencies are getting the volume, and how the rates compare to what we've negotiated. And, critically, I can show them this early enough that they can actually do something about it. Adjust the hiring plan. Redirect volume to the preferred agencies with better rates. Bring a staffing consolidation initiative forward on the roadmap.

That's the difference between backward-looking reporting and real-time intelligence. One tells you what happened. The other tells you what's happening and gives you time to shape the outcome.

Real-time visibility also changes how procurement manages its own performance. When you can watch contract compliance rates, maverick spending, and realized savings moving in real time rather than waiting for the quarterly scorecard, you can adjust your approach continuously rather than discovering problems after they've compounded for months. A sourcing initiative that isn't delivering the expected savings becomes visible in weeks, not quarters. A preferred supplier agreement that business units are ignoring shows up in the data immediately, not in the next annual review. The feedback loop between action and result tightens from months to days, and that faster feedback enables faster learning, faster correction, and faster value delivery.

Predictive Spend Analytics

This is the frontier—the capability that's emerging now and that will become standard practice within the next few years. If the capabilities I've described so far are about seeing the present more clearly, predictive analytics is about seeing the future.

Machine learning models, trained on historical spend patterns, can forecast future spending with meaningful accuracy. They can predict which categories are likely to see price increases based on commodity market trends, supplier financial health, and macroeconomic indicators. They can forecast demand patterns that affect procurement planning—predicting, for instance, that a seasonal spike in a particular category is going to be larger this year based on leading indicators in the business data. They can identify contracts that are likely to see cost overruns based on early spending patterns and flag them for proactive intervention.

This capability is still maturing, and I want to be appropriately humble about where we are. Predictive models are only as good as the data they're trained on and the assumptions built into them, and they're better at identifying trends and probabilities than at making precise point predictions. If a model tells you that there's a seventy percent probability that your logistics costs will increase by eight to twelve percent in the next quarter, that's useful intelligence for planning purposes—but it's not a guarantee, and you shouldn't treat it as one.

That said, even imperfect forecasting is significantly better than no forecasting. Most procurement organizations today are essentially flying blind into the future. They know what they spent last year, they have a budget for this year, and they hope the two numbers are somehow related. Predictive spend analytics gives them something they've never had before: an evidence-based, data-driven view of what's likely to happen, grounded in patterns that the human brain simply cannot detect across millions of transactions and dozens of external data sources.

THE CROSS-FUNCTIONAL INTELLIGENCE LAYER

Spend analysis, when done well, generates intelligence that is valuable far beyond the procurement function. The organizations that unlock this cross-functional value are the ones that get the most return from their spend analytics investment.

Finance.

CFOs and finance teams are hungry for spend data that helps them forecast more accurately, identify cost reduction opportunities, and ensure budget compliance. When procurement's spend analytics platform provides real-time visibility into committed versus actual spending by cost center, finance gets forecast accuracy that their own systems, which typically track budgets and actuals but not procurement commitments, can't provide alone. I've seen organizations where the integration of procurement spend data into financial planning processes produced meaningful improvements in forecast accuracy that reduced the painful year-end budget scrambles every CFO dreads.

Compliance and Audit.

Internal audit teams spend enormous effort testing procurement transactions for compliance with organizational policies, segregation of duties requirements, and regulatory obligations. AI-powered anomaly detection (the same capability that helps procurement identify pricing errors and maverick spending) can be configured to flag the specific patterns that internal audit cares about: transactions that bypass approval workflows, purchases that violate segregation of duties requirements, spending with sanctioned entities, and patterns consistent with fraud risk indicators. Sharing this capability with auditors doesn't dilute procurement's value. It amplifies it by demonstrating that procurement's technology investments benefit the broader organization.

Operations.

Operations teams need visibility into procurement

spending patterns to plan production, manage inventory, and optimize capacity. When spend data is connected to operational planning, and operations can see not just what was purchased but what is committed, what is in transit, and what is forecasted, then they can make better decisions about production scheduling, safety stock levels, and capacity utilization. This integration is particularly valuable in environments where input procurement timing directly affects production efficiency.

Sustainability and ESG.

Increasingly, organizations need to report on the environmental and social impact of their supply chains, including carbon emissions, labor practices, diversity spending, and environmental compliance.

Spend data is the foundation of this reporting. It tells you who you're buying from, how much, and in which categories. AI-powered spend analytics that classify and categorize supplier spending can be enriched with ESG data to provide the sustainability reporting that boards, investors, and regulators are increasingly demanding.

The procurement leader who proactively offers spend intelligence to these cross-functional stakeholders isn't just being generous with data. They're building organizational allies who understand and support procurement's technology investments because they're directly benefiting from them. This support matters when budget season arrives and the AI platform subscription needs to be renewed.

PUTTING IT INTO PRACTICE: A REAL WORKFLOW

Let me make all of this concrete with a workflow that I've actually used, because the difference between "AI can do spend analysis" and "here's how you actually do AI-powered spend analysis" is the difference between a brochure and something you can use at work tomorrow.

The scenario: you've been asked to identify the top ten spend categories with the greatest potential for sourcing savings. You have access to your organization's transaction data and a generative AI tool.

Step one: data extraction and initial assessment.

Pull ideally twelve to twenty-four months of your transaction data from your ERP or procurement system. Export it as a structured file.

Before you do anything else, upload a sample to your AI tool and ask it to assess the data quality:

- How many transactions have missing supplier names?
- What percentage have category codes?
- How many unique supplier name variations exist?
- What percentage of transactions are coded to catch-all categories?

This gives you a realistic picture of what you're working with before you start trying to analyze it. I've been surprised more than once by data that I thought was reasonably clean turning out to have significant quality issues that would have undermined the entire analysis if I hadn't checked first.

Step two: AI-assisted classification.

If your data has poor or inconsistent category coding (and it probably does) use AI to reclassify it. If you're working with a purpose-built spend analytics platform, this happens automatically. If you're working with a general-purpose AI tool, you can process the data in batches, providing the tool with your category taxonomy and asking it to classify transactions based on supplier name, transaction description, and any other attributes available. Be specific about your taxonomy. Don't just say "classify these transactions." Give the tool the actual list of categories you use, with definitions and examples of what belongs in each one. The more context you provide, the more accurate the classification will be. If you are unsure what to give it, just ask the AI. It can be more helpful that you can imagine.

The prompt I actually use for this:

"I need you to classify the following procurement transactions into our category taxonomy. Here is our taxonomy with definitions: [paste your category list with descriptions]. For each transaction, use the supplier name, transaction description, GL code, and amount to determine the most appropriate category. If you are less than 80% confident in a classification, flag it for human review. Present the results in a table showing: original data, assigned category, and confidence level (high/medium/low)."

The full version of this prompt, plus variations for different data states, is in Appendix A.

Verify a random sample of the classifications manually. *You should be spot-checking at least ten percent to validate accuracy.* Pay particular attention to transactions that the AI classified with low confidence, if the tool provides confidence scores. And look specifically at the largest dollar-value transactions that were reclassified from where they were originally coded. Those are the ones where an error would have the biggest impact on your analysis.

Step three: supplier normalization.

Ask the AI to identify likely duplicate suppliers—different name variations that appear to be the same entity. Review the suggested matches, confirm or reject them, and build your consolidated supplier view. This is where the AI saves the most human labor, because manually comparing thousands of supplier names is one of the most tedious tasks in all of procurement analytics. The AI can present you with a list of probable matches ranked by confidence, and you just need to confirm or reject each one. What used to take days, now takes hours.

Pay particular attention to cases where the AI flags a match that you wouldn't have caught—subsidiaries of the same parent company operating under completely different names, for instance. And pay equal attention to cases where the AI suggests a match that isn't actually correct—two companies with similar names that are in fact different entities. The human judgment call on borderline cases is where you earn your salary.

Step four: pattern analysis.

Now you have clean, classified, normalized spend data. This is where the analysis gets interesting. Ask the AI to identify the top spend categories by total value, the categories with the highest number of suppliers, the categories where spend is most fragmented across business units, the categories where spending has been increasing fastest, and the categories where there's the widest variation in pricing between different transactions for similar goods or services. Each of these patterns is a potential sourcing opportunity.

High spend with many suppliers suggests consolidation potential. You may be able to reduce the supply base and negotiate better terms with the remaining suppliers in exchange for higher volume. Fragmented spend across business units suggests lack of coordination which can lead to different parts of the organization buying the same things independently, probably at different prices, without leveraging the organization's combined purchasing power. Rapidly increasing spend suggests a category that may not have been strategically sourced recently, or that is experiencing market-driven cost increases that require a sourcing response. High pricing variation within a category suggests either inconsistent negotiation or a lack of benchmarking—both of which are fixable.

Step five: opportunity sizing.

For your top candidates, use the AI to help estimate savings potential based on the patterns it's seeing.

- How much spend could be consolidated under fewer suppliers?
- What's the typical range of pricing variation within the category?
- If you brought the highest-priced transactions in line with the lowest-priced transactions for comparable goods and services, what would the savings be?
- If you reduced the supplier count by fifty percent and negotiated a volume-based discount, what's a reasonable estimate of the rate improvement?

The AI's estimates here are directional, not precise, and you should treat them as such.

They're useful for prioritization: which categories are worth your team's time and energy to pursue?

They're not useful as commitment numbers: don't walk into a CFO meeting and promise the savings figure that the AI generated, because the AI doesn't know the competitive dynamics of your specific market, the switching costs involved, or the political realities that might make a category more difficult to source than it appears on paper.

You know all of that. Use the AI's estimates as a starting point, adjust them based on your professional judgment, and present a range rather than a point estimate.

Step six: deep dive on top opportunities.

For your top three to five categories, use the AI to go deeper.

- Pull the contract history—are there existing agreements, and when do they expire?
- Map the supplier landscape—who else is in this market, and what do you know about their capabilities and pricing?
- Identify the key stakeholders—which business units are the primary users, and who would need to be involved in a sourcing initiative?
- Assess the risk factors—is this a category where supply disruption would be particularly painful, and if so, does that affect how aggressively you should consolidate?

Each of these questions would traditionally require separate research efforts. With AI assistance, you can build a comprehensive category profile in hours rather than weeks—a profile that gives you the intelligence you need to develop a sourcing strategy and make a compelling case to stakeholders and leadership.

Step seven: stakeholder presentation.

Use the AI to help you build the narrative around the data. Ask it to generate an executive summary of the key findings, highlighting the most significant opportunities and the recommended prioritization. Then—and this is critical—***apply your own judgment***. The AI doesn't know that Category X is politically sensitive because the CFO's former colleague runs the incumbent supplier. It doesn't know that Category Y has an expiring contract in sixty days, which makes it a more urgent opportunity than the larger Category Z. It doesn't know that your organization tried to consolidate Category W three years ago and the initiative failed because the business units wouldn't cooperate. You know all of that. Layer it in. The AI provides the analytical foundation. You provide the organizational intelligence that transforms analysis into actionable strategy.

WHERE TO START: PRIORITIZING THE ADVANCED APPLICATIONS

Once you have the basic AI-powered spend visibility working, three advanced applications open up—and the question of which one to tackle first matters more than the question of which is most exciting.

The honest answer is: it depends on your organization.

If your organization has lots of negotiated contracts but loose enforcement of preferred-supplier agreements, start with contract compliance monitoring. The savings are concrete, the time-to-value is fast, and the visibility you create makes the next two applications easier.

If your organization has a long tail of small purchases that has historically been ignored because the economics didn't justify strategic sourcing attention, start with tail spend management. The volume of opportunity is usually larger than people expect, and the early wins build credibility for further investment.

If your organization is preparing to refresh sourcing strategies for a major category (and most organizations are, on a rolling basis) use AI-powered spend analytics as the input to category strategy development. That's the highest-leverage use; it's also the one that most directly demonstrates AI's strategic value to senior leadership.

None of these is wrong. The right order is the one that aligns with your organization's most immediate need and produces a visible result your stakeholders will recognize.

BEYOND THE BASICS: ADVANCED APPLICATIONS

Once you have solid AI-powered spend visibility in place, a set of more sophisticated applications becomes possible—applications that most organizations haven't even considered because they didn't have the data foundation to support them.

Contract Compliance Monitoring

One of the most persistent sources of value leakage in procurement is contract non-compliance—transactions that should be flowing through a negotiated agreement but are instead being purchased off-contract, usually at higher prices. In most organizations, compliance monitoring is a periodic exercise: someone pulls a report, compares transactions against contracted rates and preferred suppliers, identifies the gaps, and then tries to get the business to change its behavior. It's reactive, it's slow, and it's usually incomplete because the spend data isn't clean enough to make the comparison accurate.

With AI-powered spend classification and real-time visibility, compliance monitoring can become continuous and automated. The system flags off-contract purchases as they happen and can automatically route alerts to the appropriate category manager or business unit leader. This changes the dynamic from periodic compliance policing to real-time nudging, and the difference in outcomes is significant. Organizations that implement real-time compliance monitoring typically see contract compliance rates improve by ten to twenty percentage points within the first year, which translates directly to savings because contracted rates are almost always better than spot-buy rates.

Tail Spend Management

Tail spend, the long tail of small-dollar purchases that collectively represent a significant amount of money but are individually too small to justify strategic sourcing attention, is one of the most persistently frustrating challenges in procurement. It typically represents fifteen to twenty percent of total spend but involves sixty to eighty percent of suppliers and transactions. The economics of traditional strategic sourcing don't work for tail spend: you can't justify a six-month sourcing project for a category where the total spend is $50,000.

AI changes this calculus. When you can classify and analyze tail spend automatically, you can identify patterns that were previously invisible. Maybe those fifty small purchases of office furniture from twenty different suppliers could be consolidated into a single preferred supplier program. Maybe those hundreds of small consulting engagements could be channeled through a managed services arrangement. The individual transactions aren't worth sourcing strategically, but analyzing the full tail at once often reveals opportunities that are.

I've seen organizations capture five to fifteen percent savings on their tail spend through AI-enabled analysis and consolidation—and given that tail spend at a large organization can represent hundreds of millions of dollars, those percentage points add up quickly.

Category Strategy Development

The most sophisticated application of AI-powered spend analytics is as an input to category strategy development. When you have comprehensive, accurate, real-time spend data, enriched with supplier intelligence and market data, you can develop category strategies that are grounded in evidence rather than intuition.

What does this look like in practice? Instead of developing a category strategy based on last year's spend report and the category manager's general knowledge of the market, you develop it based on a real-time view of spending patterns, supplier performance data, market pricing intelligence, contract expiration timelines, risk assessments, and demand forecasts—all integrated, all current, all available at the click of a button rather than after weeks of manual research.

The category manager's expertise doesn't become less important in this model. It becomes more leveraged. Instead of spending sixty percent of their time collecting and cleaning data and forty percent analyzing it and developing strategy, they spend ten percent on data, because the AI handles the rest, and ninety percent on analysis, strategy, and stakeholder engagement. The quality of their thinking improves because they have better inputs. The speed of their execution improves because they're not waiting for data. And the credibility of their recommendations improves because they're backed by comprehensive, current, verifiable data rather than approximations and estimates.

THE TRAPS: WHERE AI SPEND ANALYSIS GOES WRONG

Because I promised you honesty alongside enthusiasm, let me walk you through the most common ways AI-powered spend analysis goes sideways. I've seen all of them, and some of them are costly.

The garbage-in-garbage-out trap.

AI cannot magically fix fundamentally broken data. If your transaction data is missing critical fields (no supplier names, no descriptions, no meaningful coding of any kind) even the most sophisticated AI model will struggle to produce useful classification. AI significantly reduces the manual effort required to clean and classify data, but it doesn't eliminate the need for reasonable data quality at the input stage. If your data is truly terrible—and I mean "we literally don't know who we paid or what we paid them for" terrible—the first step isn't AI. It's fixing your data capture processes at the source. Talk to your AP team. Talk to your ERP administrators. Get the fundamentals of data capture right, and then let AI do the heavy lifting on classification and analysis.

The false precision trap.

AI-powered spend analysis can produce output that looks remarkably precise and authoritative, containing beautiful dashboards, clean categories, specific dollar amounts down to the penny. This precision can create a false sense of confidence in the underlying accuracy. Always remember that the classification is probabilistic, not certain. A ninety-three percent accuracy rate means that seven percent of your transactions are in the wrong category. On a billion-dollar spend base, that's $70 million of misclassified spend. That's not a rounding error. It's a significant uncertainty that needs to be acknowledged in how you present and use the data.

This is particularly important when you're using spend data to build business cases for sourcing initiatives. If your AI-powered analysis shows $50 million in a category, and you build a sourcing strategy that promises ten percent savings, $5 million, but the actual category spend is only $42 million because seven percent was misclassified, your savings target was wrong from day one. Verify the numbers in your highest-priority categories. Spot-check the classifications. Maintain appropriate humility about the limitations of any automated classification. And present ranges rather than point estimates when the stakes are high.

The one-and-done trap.

Spend analysis is not a project. It's a practice. The organizations that get the most value from AI-powered spend analytics are the ones that treat it as a continuous capability—not a one-time exercise that produces a report that goes on a shelf. The AI model needs to be monitored, its accuracy needs to be validated regularly, and its classifications need to be updated as your supplier base and category taxonomy evolve. New suppliers appear. Categories get restructured. Business units reorganize. Mergers and acquisitions bring new systems and new data. A model that was ninety-three percent accurate last year might be eighty-five percent accurate this year if it hasn't been updated to reflect changes in the environment. Set it and forget it is not a strategy. Treat the model like any other critical business tool: maintain it, validate it, and invest in keeping it current.

The insight-without-action trap.

This one isn't an AI problem—it's a procurement problem that AI can actually make worse. Better spend visibility produces more insights. More insights demand more action. If your organization doesn't have the capacity, the mandate, or the stakeholder support to act on the sourcing opportunities that AI-powered spend analysis reveals, you're going to produce a lot of very pretty dashboards that don't actually change anything.

I've seen this pattern more times than I'd like to admit. A procurement team invests in AI-powered spend analytics, discovers significant savings opportunities, gets excited about the potential—and then runs into the organizational reality that acting on those opportunities requires stakeholder alignment, cross-functional collaboration, and executive sponsorship that the procurement team doesn't have. The insights sit in a dashboard. The savings remain theoretical. And eventually, someone asks why the organization invested in fancy analytics tools that haven't delivered measurable results.

The value of spend analysis, AI-powered or otherwise, is not in the analysis itself. It's in the decisions and actions that the analysis enables. Make sure you have the organizational capacity and mandate to act before you invest heavily in the capability to see. Or, at minimum, plan your implementation to target a specific, actionable use case first (a single category with clear savings potential, an engaged stakeholder, and a realistic execution timeline) so you can demonstrate value before you scale.

THE COMPETITIVE ADVANTAGE: WHAT GREAT LOOKS LIKE

Let me close this chapter by painting a picture of what AI-powered spend analysis looks like when an organization is doing it well—because the gap between "doing it" and "doing it well" is significant, and the competitive advantage lives in the latter.

The organizations that are leading in this space have continuous, near-real-time visibility into their procurement spending across the entire enterprise. Their spend data is automatically classified, their suppliers are automatically normalized, and their dashboards reflect what happened today, not what happened last quarter. Their procurement teams spend almost no time on the mechanics of data collection and classification, that's handled by the AI, and almost all of their time on analysis, strategy, and action. They can identify a sourcing opportunity on Monday, build the business case on Tuesday, and begin executing on Wednesday—a cycle that used to take months.

Their category managers have a depth of spend intelligence that was previously available only to the largest, most well-resourced procurement organizations. They know not just how much they're spending in a category, but how that spend is distributed across suppliers, business units, geographies, and contract vehicles. They know where they're compliant with preferred supplier agreements and where they're not. They know which categories are trending up and which are trending down. They know where their pricing compares favorably to market and where it doesn't. They can identify patterns across categories (that the same supplier is creeping up in multiple categories, or that a particular business unit is consistently buying outside of negotiated channels) that would be invisible in siloed, category-by-category analysis.

And perhaps most importantly, they're using that intelligence to have fundamentally different conversations with their business stakeholders. Not "here's how much we spent last year"—that's reporting, not insight. But "here's what I'm seeing in your division's technology spending that suggests we have a $2 million consolidation opportunity, and here's a sourcing strategy that could capture it without disrupting your operations." That's the kind of conversation that transforms procurement from a back-office function into a strategic advisory partner. And it's the kind of conversation that's only possible when you have the spend visibility to back it up.

The AI didn't create the opportunity. The opportunity was always there—buried in the data, hidden by inconsistent coding, obscured by fragmented systems, invisible to the quarterly report. What AI did was make the invisible visible. What the procurement professional did, what only the procurement professional can do, was to see the opportunity in the data, build the strategy around it, navigate the organizational dynamics required to execute it, and make it happen.

That's the model. Not AI replacing human judgment. AI revealing the landscape. Human judgment navigating it.

The Model: AI revealing the landscape. Human judgment navigating it.

Now let's apply this to the intelligence you need before you ever sit down at a negotiation table or launch a sourcing event. Let's talk about market research and supplier intelligence at machine speed.

CHAPTER 5
MARKET INTELLIGENCE AT MACHINE SPEED

KNOWING MORE THAN THE OTHER SIDE OF THE TABLE

There's a moment in every significant procurement negotiation, whether it be a sourcing event, supplier selection, or contract renewal, where the outcome hinges on who knows more. Not who's tougher. Not who's more aggressive. Not who has the better poker face. Who knows more.

The buyer who understands the supply market's pricing dynamics, the supplier's financial pressures, the competitive landscape's recent shifts, and the emerging alternatives that didn't exist eighteen months ago walks into the room with options that the less-informed buyer doesn't have. The buyer who's running on last year's benchmarks, a vague sense that "the market has been competitive," and a hope that the supplier's opening position has some room in it is bringing a knife to a gunfight. A butter knife, at that.

I have been on both sides of this information gap over the course of my career, and I can tell you with absolute certainty which side is more fun.

Let me tell you about a story I now use in my Rutgers classes, because it captures a lesson every procurement professional needs to learn, and most of us learn it the hard way.

A colleague of mine, early in his career, walked into a negotiation with a major packaging supplier armed with what he thought was solid market intelligence. He had pricing data from the last competitive bid the company had run two years earlier. He had a general sense from trade publications that the raw material costs in the category had been declining. And he had the confident assumption that since the market was moving in his favor, he could push hard on price and the supplier would have to accommodate because their costs were going down.

The supplier's lead negotiator, a veteran who had probably forgotten more about packaging markets than my colleague had ever learned, calmly walked him through a fifteen-minute presentation that showed exactly how wrong his assumptions were. Yes, raw material prices had declined over the previous twelve months. But they had rebounded sharply in the most recent quarter due to a supply disruption upstream. Yes, there were alternative suppliers in the market, but two of the three he might have considered had been acquired by competitors in the past six months, significantly consolidating the supply base. And the volume projections he was using as leverage? The supplier knew (because they tracked actual orders, not just forecasts) that volumes had been declining for three consecutive quarters, which meant his leverage was weaker than he was pretending it was.

He got educated that day. Expensively. And the lesson that's now baked into every class I teach and every preparation framework in this book wasn't "be tougher." It was "be better informed." Information asymmetry is the single greatest determinant of negotiation outcomes, and the party with the superior information almost always gets the superior result.

As we've discussed, for most of the history of the procurement profession, building that superior information position required enormous effort. Days or weeks of research. Expensive analyst reports and market intelligence subscriptions. Conversations with industry contacts who might or might not have current information. Internal data that was scattered across systems and required significant effort to compile and interpret. The information was theoretically available, but the time and cost of assembling it meant that thorough market intelligence was reserved for the biggest, most strategic sourcing events—and even then, it was often incomplete.

Imagine Big Ron with this kind of market intelligence.

For his top fifteen suppliers, he didn't need much of it. He had been doing the job long enough that he knew the players, the pressures, the recent leadership shuffles. He knew which sales VP had a number to make and which one had just been passed over for a promotion. He knew which suppliers had lost a major account because somebody had called somebody at the funeral of somebody else's father. Big Ron's intelligence on the top of his portfolio was, in its way, excellent—built through three decades of dinners, industry conferences, and the fragments of overheard information that flow to operators who've been in the field that long.

For the other several hundred suppliers in his portfolio, he had nothing. He walked into those renewals with last year's pricing, a vague sense of the market, and whatever the supplier's account rep volunteered in the first ten minutes of the meeting. He couldn't help it. Real intelligence on every supplier scaled linearly with headcount, and Big Ron didn't have the headcount.

Give Big Ron the kind of market intelligence this chapter describes—financial signal extraction from earnings calls in forty minutes, supplier-by-supplier pressure-point mapping in an afternoon, alternative-supplier landscapes refreshed quarterly—and the portfolio that used to be invisible would have opened up. Not as a change of character. Big Ron wasn't going to start sending Christmas cards because his dossiers got better. As an operator, though, he would have walked into the renewals at the tail of his portfolio knowing more than the supplier expected him to know. What he did with that knowledge—pressure or partnership—was always going to be Big Ron's call. Same as it'll be yours.

That constraint has been obliterated. AI has not changed what market intelligence is or why it matters—those are constants that will endure for as long as markets exist. What AI has changed is how fast you can build it, how comprehensive it can be, and who has access to it.

This chapter is about exactly how to do that. Not theoretically. Practically—with workflows, specific approaches, and specific examples of what great AI-powered market intelligence looks like and how it translates into better procurement outcomes.

Senior leaders: Here's the strategic argument before we get into the procurement professional mechanics.

The historical information asymmetry between procurement organizations and their counterparties, sales teams that knew their market deeply versus buyers who had time to research only their largest deals, is reversing. The well-resourced sales organization no longer has a structural information advantage over the procurement professional who knows how to direct AI tools. That reversal has consequences that compound over every supplier relationship a procurement function manages. The strategic question is not whether market intelligence matters. It's whether your function is going to capture the asymmetry reversal as a structural competitive advantage or watch competitors capture it. The chapter that follows is the playbook for the first option.

THE INTELLIGENCE YOU NEED: A FRAMEWORK

Before we talk about how to gather intelligence, let's be precise about what intelligence you actually need. Because one of the traps of having a powerful research tool is the temptation to research everything—to go so deep and so broad that you end up with a hundred pages of analysis and no clear picture of what it means for your decision.

Great procurement market intelligence is organized around four questions. Everything else is noise.

Question One: What Does the Supply Market Look Like?

This is the foundational question—the lay of the land.

- How many qualified suppliers exist in this category?
- How is market share distributed among them?
- Is the market consolidating or fragmenting?
- Are there new entrants that are changing the competitive dynamics?
- What's the capacity utilization in the market—is supply tight or abundant?
- What are the barriers to entry that might prevent new competitors from emerging?

The answers to these questions define your structural leverage. A market with many qualified suppliers and abundant capacity is a market where you have options and where competitive bidding will generate pressure. A market dominated by two or three large players with high barriers to entry is a market where your leverage is constrained by structure, and where creative negotiation strategies become more important than competitive pressure.

Let me give you an example of how AI transforms this analysis.

Last year, I was preparing a sourcing strategy for a specialized professional services category—the kind of niche consulting that has maybe twenty credible providers globally. In the old world, mapping this market would have required a combination of industry analyst reports, referral calls to colleagues at other companies, LinkedIn searches, and conference networking—probably two to three weeks of elapsed time to build a reasonably complete picture.

Using AI tools, I built the same market map in about four hours. I asked the AI to identify the major providers in the category, then to dig into each one's recent activity: leadership changes, client wins and losses, geographic expansion, M&A activity. I asked it to synthesize industry analyst commentary on competitive dynamics and emerging trends. I asked it to identify boutique providers and emerging entrants that might not show up in traditional analyst coverage. The output was a comprehensive market landscape that gave me a strong starting position for stakeholder conversations and supplier outreach.

Was it perfect? No.

The AI missed two boutique providers that I learned about through a conversation with an industry contact two days later. Was it good enough to dramatically accelerate my market understanding and direct my subsequent research?

Absolutely.

Most of the negotiations I've watched go badly went badly here. Not in the room—in the prep. The buyer didn't actually know what was driving the supplier's cost, so the conversation in the room turned into an argument about positions instead of a conversation about reality. The supplier always wins that argument because they live inside their cost structure all day, every day.

Question Two: What's Driving Pricing in This Market?

Understanding price drivers, the underlying cost structure and market dynamics that determine where prices are headed, is the difference between negotiating from informed positions and negotiating from positional posturing.

What are the major cost components in this category? How have those cost components moved over the past twelve to eighteen months? What's the forecast for those components going forward? What macro factors (currency movements, regulatory changes, supply disruptions, demand shifts) are affecting pricing? Is the market in a buyer's phase (excess capacity, declining demand, suppliers fighting for share) or a seller's phase (constrained supply, strong demand, suppliers with pricing power)?

These are questions that AI can help you answer with remarkable speed. Commodity price indices, raw material cost trends, labor market data, energy costs, shipping rates, currency movements—all of this information is publicly available, and generative AI tools are excellent at synthesizing it into a coherent picture of what's actually happening in the cost environment.

I used this capability recently in a negotiation with a martech platform vendor who was pushing for a twelve percent price increase, citing "unprecedented cost pressures across the industry." Before the negotiation, I asked an AI tool to analyze the key cost drivers in this category over the previous twelve months—cloud infrastructure costs, engineering talent compensation in the vendor's primary geography, third-party data licensing fees, and overall SaaS pricing trends in the segment. The analysis took about forty-five minutes and produced a detailed picture that told a very different story from the vendor's narrative.

Cloud infrastructure costs had indeed increased—but by about six percent, not twelve, and the largest provider in the segment had just announced price reductions for new commitments. Engineering compensation in the vendor's geography had stabilized after a hiring frenzy earlier in the year. Third-party data licensing fees had ticked up modestly. And the broader martech segment was showing pricing softness, not pricing pressure—competitors were offering aggressive renewals to retain customers in a tightening enterprise software budget environment.

Armed with this analysis, I could walk into the negotiation and have a different conversation. Not "your increase is too high"—which is a positional statement that invites an adversarial back-and-forth. But "we've analyzed the cost drivers in this category and we see feedstock up about six percent, energy flat, transportation down, and labor up moderately. Can you walk us through which specific cost elements are driving the additional six points in your proposed increase beyond what the market data suggests?" That specific, evidence-based, curious rather than accusatory question completely changed the dynamic. The supplier's team acknowledged that their initial proposal had included some margin recovery, and we ultimately agreed on an increase of about five percent, less than half their opening position, supported by a shared analysis of actual cost movements.

That's the power of better information applied with the preparation discipline and the curiosity we discussed in Chapter Two.

Question Three: What's Going On Inside This Specific Supplier?

General market intelligence is valuable. Supplier-intelligence is transformative. Understanding what's happening inside the specific company you're about to negotiate with (their financial health, their strategic priorities, their recent wins and losses, their organizational changes, their capacity constraints) gives you the ability to customize your negotiation approach to their specific situation rather than applying a generic strategy.

This is where AI-powered research becomes almost unfairly powerful. For publicly traded suppliers, the amount of information that AI can synthesize is hard to overstate.

- Earnings call transcripts, where executives discuss their strategic priorities, margin pressures, and growth plans.
- Annual reports and SEC filings, where financial performance, risk factors, and strategic direction are documented in detail.
- Analyst reports, where independent researchers

assess the company's competitive position and future prospects.

- Press releases, where new product launches, partnership announcements, and organizational changes are documented.
- LinkedIn, where changes in the account team, leadership transitions, and hiring patterns can signal strategic shifts.
- Industry publications, where the supplier's market activities and competitive moves are tracked.

The prompt I use to build a supplier financial dossier: "Analyze [Company Name]'s financial performance and strategic position as a potential/current supplier to my organization. Based on their most recent public filings and earnings communications, provide: (1) Revenue trajectory—growth rate over past 3 years, and current guidance. (2) Profitability—gross margin and operating margin trends, including any margin pressure or expansion commentary from management. (3) The business segment that would serve our account—its specific performance, growth dynamics, and competitive position. (4) Recent strategic announcements, M&A activity, or leadership changes. (5) Any disclosed risks that could affect their ability to serve us reliably. Cite your sources for each point." Full prompt with private-company variant in Appendix A.

With AI, you can build a comprehensive supplier dossier in an hour. The process looks something like this:

1. Upload the supplier's most recent earnings call transcript and ask the AI to identify every mention of margin pressure, pricing strategy, capacity investment, customer retention, and competitive threats.
2. Then upload or reference their most recent annual report and ask for a summary of financial performance trends—revenue growth, margin trajectory, debt levels, cash flow, and any risk factors that could affect their ability to perform on your contract.
3. Ask the AI to search for recent news about the company—leadership changes, major contract wins or losses, legal issues, regulatory actions, acquisition activity.
4. And ask it to analyze the LinkedIn profiles of the people who will be in the negotiation—how long they've been in their roles, what their background is, what their likely incentive structure looks like based on their title and the company's compensation practices.

The logistics negotiation I described in Chapter Two—where the AI surfaced that the supplier had lost two major accounts in our region—is a perfect example of this kind of supplier-intelligence changing a negotiation outcome. Those two account losses were not information that the supplier volunteered. They were not information that appeared in any single source. They were fragments scattered across a trade publication, a LinkedIn post, and an analyst Q&A that, individually, would have been invisible to me. The AI connected the dots because that's what AI does well—synthesizing signals across multiple sources into a coherent pattern that informs a decision.

And let me be clear about what I did with that intelligence in the negotiation. I didn't wave it in the supplier's face. I didn't say "we know you lost two accounts, so you're desperate." That would have been the Big Ron move—using intelligence as a weapon. Instead, I used it to inform my understanding of their situation. I knew they had more motivation to retain our business than their confident opening position suggested. That knowledge gave me the confidence to push harder on price than I might have otherwise, and the patience to hold my position when they pushed back, because I understood that their alternatives were weaker than they were projecting.

That's the difference between using intelligence as a weapon and using intelligence as insight. Both give you advantage—but one preserves the relationship and the other damages it. Chapter Two's lessons about principled negotiation and ethical practice apply here with full force.

Question Four: What Are My Alternatives?

The fourth intelligence question connects directly to BATNA—the best alternative to a negotiated agreement that we discussed in Chapter Two. Building a strong BATNA requires you to know what alternatives actually exist in the market, how capable they are, what they would cost, and how feasible a transition would be.

AI accelerates alternative identification significantly. If

you're locked into a single-source technology provider and you want to understand what your options actually look like, you can ask an AI tool to identify alternative providers, compare their capabilities against your requirements, estimate the transition effort and cost, and assess their track record with organizations similar to yours. The analysis won't be definitive, you'll still need to engage with potential alternatives directly to validate fit and feasibility, but it gives you a map of the landscape that allows you to assess whether your BATNA is "no alternative" or "alternatives I haven't investigated yet."

I've used this capability to challenge assumptions about sole-source categories that turned out to be less sole-source than everyone believed. In one case, a technology platform that the organization had been treating as irreplaceable for five years turned out to have two credible competitors that had entered the market in the previous eighteen months that nobody in the organization had evaluated because "we're sole-sourced" had become an unchallenged assumption rather than a regularly validated fact.

The AI identified both competitors in about twenty minutes. The subsequent evaluation took several weeks and ultimately confirmed that one of the two was a viable alternative. That alternative became a credible BATNA that reshaped the dynamics of the next contract renewal, generating concessions that would never have been possible if the "sole-source" assumption had gone unchallenged.

THE WORKFLOW: BUILDING A SUPPLIER INTELLIGENCE DOSSIER IN NINETY MINUTES

Let me walk you through the specific workflow I use to build a comprehensive supplier intelligence dossier using AI tools. This is the practical, step-by-step process that turns the conceptual framework we just discussed into an actual preparation document you can use before a negotiation or sourcing event.

I'm going to use a realistic scenario: you're preparing for a contract renewal with a mid-market IT services provider. The contract is worth $2.5 million annually, the current term expires in four months, and the supplier has signaled that they'll be seeking a price increase at renewal. Your goal is to walk into the negotiation with a comprehensive understanding of the market, the supplier, and your options.

Step One: Market Context (20 minutes)

Start broad.

Ask the AI to give you a current overview of the IT services market segment this supplier operates in.

1. What's the overall market growth rate?
2. How is the competitive landscape structured?
3. What are the major trends affecting pricing—labor cost inflation, offshore delivery model

shifts, automation, AI's impact on service delivery economics?
4. What's the supply-demand balance—is it a buyer's market or a seller's market?

Then narrow.
Ask for specifics about the mid-market segment.

1. How does pricing in this segment compare to the enterprise segment?
2. Are there pricing pressures specific to mid-market providers that differ from the large systems integrators?
3. What's the typical margin profile for companies in this segment?

The AI will produce a market overview that you should treat as directional, not definitive. Cross-reference the major claims against one or two industry sources—a Gartner or Forrester report, an industry association publication, or a recent article from a credible trade publication. This takes about five additional minutes and gives you confidence that the AI's market characterization is reasonable.

Output: A two-to-three page market context summary that you understand well enough to reference in conversation—not to present formally, but to inform your intuition about what's reasonable.

Step Two: Supplier Financial Analysis (25 minutes)

Publicly Traded:

If the supplier is publicly traded, this step is rich. Upload their most recent earnings call transcript—these are freely available from the company's investor relations page or from financial sites. Ask the AI to extract and summarize every mention of: revenue growth and guidance, margin trends and targets, pricing strategy, client retention and expansion metrics, capacity and hiring plans, competitive positioning comments, and any discussion of challenges or headwinds.

Then pull their most recent 10-K or annual report. Ask the AI to identify: revenue trends over the past three years, gross and operating margin trajectory, customer concentration (are they dependent on a few large clients?), debt levels and cash flow position, and any risk factors disclosed that are relevant to their ability to serve you.

Privately Held:

If the supplier is privately held, which is common in the mid-market IT services space, you'll have less financial data to work with, but the AI can still help. Ask it to estimate the company's revenue range based on publicly available indicators—employee count (from LinkedIn), office locations, client list (from their website and case studies), and any press mentions of contract values. The estimates will be rough, but even a directional understanding of a supplier's size and trajectory is better than no understanding at all.

What you're looking for: Signals that inform your

negotiation strategy. Is this supplier growing rapidly and confident, or struggling and defensive? Are their margins expanding (suggesting room to negotiate) or contracting (suggesting cost pressure)? Are they investing in automation and AI that might reduce their delivery costs—and if so, should those cost reductions be reflected in your pricing? Is there anything in their financial profile that suggests vulnerability—high debt, customer concentration, declining win rates—that would increase the value of retaining your business?

Output: A one-to-two page financial summary with the key data points highlighted and your interpretation of what they mean for the negotiation.

Step Three: Supplier-Specific Intelligence (20 minutes)

This is where you go beyond the financials into the softer intelligence that often proves most valuable in negotiations. Ask the AI to research recent news about the supplier—leadership changes, major contract wins or losses, strategic pivots, partnership announcements, awards and recognition, or negative press. Ask it to analyze the LinkedIn profiles of the people who will be at the negotiation table—their tenure, their background, their likely priorities.

Then ask the AI a question that most procurement pros don't think to ask: "Based on the information you've analyzed about this company, what are the three things that are most likely keeping their leadership team awake at night?" The AI's answer will be speculative, but it's usually grounded in the financial and strategic signals it's identified, and it gives you a framework for thinking about what the supplier actually cares about—not just what they're going to ask for.

I've found that this question, what keeps their leadership up at night, consistently produces the most strategically useful output in the entire dossier process. Because negotiation leverage isn't just about your alternatives. It's about understanding the other party's pressures and priorities well enough to propose solutions that address their concerns while advancing your interests.

Output: A one-page intelligence brief covering recent developments, key personnel, and strategic pressure points.

Step Four: Competitive Alternatives (15 minutes)

Ask the AI to identify the most credible alternative providers for the services you're currently buying from this supplier. For each alternative, ask for: their approximate size and market position, their key differentiators, their typical pricing model and range (if available), their reputation for quality and reliability, and any known wins or losses against your current supplier.

Then ask the AI to assess the feasibility of transitioning to an alternative: what would be involved, how long would it take, what are the risks, and what would the approximate cost of transition be? This analysis won't replace a proper evaluation, you'd need to engage with the alternatives directly for that, but it gives you a realistic picture of your BATNA that prevents you from either overestimating your alternatives (which leads to overplaying your hand) or underestimating them (which leads to leaving concessions on the table).

What you're looking for: Not necessarily a commitment to switch suppliers. Often, the most valuable outcome of the alternatives analysis is the credible knowledge that alternatives exist. That knowledge changes your psychological posture in the negotiation—you're prepared to walk away if the deal doesn't work, and that preparation shows in ways that affect the supplier's calculus even if you never explicitly reference it.

Output: A competitive alternatives summary with enough detail to inform your BATNA assessment.

Step Five: Synthesis and Strategy (10 minutes)

Now ask the AI to do something that demonstrates the real power of these tools: synthesize.

"Based on the market context, supplier financial analysis, supplier intelligence, and competitive alternatives we've analyzed, what are the five most important insights that should inform my negotiation strategy? And what negotiation approach would you recommend, including specific asks, likely supplier responses, and creative options for creating mutual value?"

The synthesis won't be perfect. The AI doesn't know your internal politics, your relationship history, or your risk tolerance. But it provides a structured starting point for your strategy that incorporates all the intelligence you've gathered—a starting point that you then refine with your own judgment, organizational knowledge, and professional experience.

Output: A negotiation strategy outline with prioritized objectives, anticipated supplier positions, and creative options.

Total time elapsed: approximately ninety minutes.

Compare that to the traditional approach where the same quality of intelligence gathering would have taken a research team one to two weeks of elapsed time, or more realistically, would simply not have been done at all because nobody had the bandwidth. The magnitude of the transformation becomes clear.

A note on what you're about to read: don't copy these. Read them, understand why they're structured the way they are, then write your own for your specific situation. A prompt copy-pasted into a context it wasn't built for produces output that's confidently wrong—which is the worst kind of wrong, because confident wrong is the version you don't catch until you're sitting in the negotiation room.

THE PROMPTS THAT ACTUALLY WORK: A PRACTICAL LIBRARY

Let me share some of the prompts I've developed and refined through months of practice. These aren't templates you should copy verbatim—they should be adapted to your specific situation, category, and supplier. But they illustrate the level of specificity and context that produces useful output versus the generic responses that make people conclude AI isn't helpful.

For market landscape analysis:

"I'm a senior procurement professional preparing a sourcing strategy for [category]. Give me a comprehensive overview of the current supply market, including: the major providers by market share and geographic presence, recent consolidation activity (M&A, partnerships, market exits), the pricing trends over the past 12-18 months and the key cost drivers, and any emerging disruptors or new entrants that are changing the competitive dynamics. Focus specifically on the [sub-segment/geography/scale] that's most relevant to a buyer spending approximately [$X] annually."

For supplier financial intelligence:

"Analyze the attached earnings call transcript from [Company]. Extract and summarize every discussion point related to: margin pressures and profitability targets, pricing strategy and customer retention, capacity constraints or expansion plans, competitive threats they're responding to, and any strategic initiatives that might affect how they approach contract renewals with existing customers. Highlight the three insights that would be most valuable for a procurement professional preparing to negotiate a contract renewal with this company."

For cost driver analysis:

"I'm preparing to challenge a [X]% price increase proposed by a supplier in the [category] space. Help me build a should-cost model by analyzing the key cost components in this category: [list relevant inputs—raw materials, labor, energy, transportation, etc.]. For each component, what has the price trajectory been over the past 12 months? What's the forecast for the next 6-12 months? And based on this analysis, what would a reasonable overall cost change be for a product/service in this category?"

For competitive alternative assessment:

"I'm currently single-sourced with [type of supplier] and need to evaluate whether credible alternatives exist. Identify the top 5-7 providers in this space that could serve a [size/industry/geography] client with requirements similar to mine. For each one, provide: their approximate scale, their key differentiators, their typical pricing model, and any publicly available intelligence about their win rate against incumbent providers. Then assess the realistic effort and risk involved in transitioning from an established provider to one of these alternatives."

For pre-negotiation synthesis:

"Act as a senior procurement strategy consultant. Based on everything we've discussed about [supplier] (their financial position, the market dynamics, the competitive alternatives, and the cost driver analysis) develop a negotiation preparation brief. Include: (1) our three strongest leverage points, (2) the supplier's likely opening position and the arguments they'll use to justify it, (3) our recommended opening position with market data to support it, (4) three creative deal structure options that could create value for both parties, and (5) the two or three issues where we should be prepared to make concessions and what we should get in return."

Each of these prompts reflects procurement expertise. They ask the questions that an experienced procurement pro knows matter, in the language that produces actionable output. A technology person without procurement experience wouldn't know to ask most of these questions. A procurement person without AI experience wouldn't know how to structure them for maximum effectiveness.

The intersection, procurement expertise plus prompting skill, is where the value lives.

WHERE THE AI GETS IT WRONG: INTELLIGENCE VALIDATION

Let me be rigorous about this, because the hallucination risk we discussed in Chapter Three is particularly dangerous in the context of market intelligence. When you cite a market benchmark or a supplier financial metric in a negotiation, you are staking your professional credibility on the accuracy of that information. If it turns out to be fabricated, the damage to your credibility can take months or years to repair.

Here are the specific validation practices I follow for every AI-generated intelligence dossier:

Financial data gets verified against primary sources. If the AI tells me a supplier's gross margin is forty-two percent, I verify it against their actual SEC filing or annual report. This takes about three minutes per data point and is non-negotiable for any number I might cite in a negotiation.

Market benchmarks get cross-referenced. If the AI provides a market pricing benchmark, I check it against at least one independent source—an analyst report, a publicly available index, or a credible industry publication. If I can't find independent corroboration, I don't use the number. I may still use the directional insight, pointing out "market pricing in this category has been declining," but I won't cite a specific benchmark I can't verify.

Recent events get confirmed. If the AI tells me a competitor was acquired or a supplier lost a major account, I verify it through a quick web search. AI tools can generate plausible-sounding events that didn't actually happen, and citing a fictional acquisition in a negotiation would be catastrophic.

Estimates get labeled as estimates. When the AI provides estimates for private company revenue, market share, or other data points that aren't publicly reported, I note them as estimates in my dossier and treat them accordingly in conversation. "Our research suggests the company is in the $200-300 million revenue range" is honest. "The company's revenue is $247 million" is false precision that will backfire if challenged.

The validation step typically adds about thirty minutes to the dossier process. That takes our total from ninety minutes to two hours. Still significantly faster than the traditional approach, and now with a level of confidence in the accuracy that lets you use the intelligence in high-stakes conversations without risk. Don't skip this step due to lack of time. Always remember how long it used to take. Cutting corners at this point in the process could make or break a deal.

FROM INTELLIGENCE TO STRATEGY: CONNECTING THE DOTS

Having great market intelligence is necessary but not sufficient. The intelligence has to be translated into specific negotiation approaches, sourcing decisions, and relationship management actions that create value. And this translation of intelligence into strategy is where procurement expertise becomes the irreplaceable ingredient.

Let me walk you through how intelligence becomes strategy with a real example.

The scenario: In a contract renewal with a major HR technology platform, the AI-powered intelligence dossier revealed the following:

The market was in transition. Two major players had merged in the previous year, reducing competitive options. The supplier's recent earnings calls showed that customer retention was their top strategic priority, with their CEO explicitly stating that reducing churn was more important than new customer acquisition in the current fiscal year. The supplier had just launched a new analytics module that they were eager to get adoption on. Their stock price had declined fifteen percent in the previous quarter following a disappointing earnings report, putting pressure on the leadership team to show improved financial results. And two alternative providers had recently enhanced their platforms to the point where migration, while still costly, was more feasible than it had been two years earlier.

Each of these intelligence points pointed toward a specific strategic move, and the picture that emerged when I connected them was clearer than any single insight would have been on its own.

The market consolidation told me competitive pressure alone wouldn't drive significant concessions. The two-player market meant we needed to be creative rather than just competitive, leveraging trade structure rather than threatening a bid we couldn't credibly run at scale. The supplier's documented retention focus shifted my read of the leverage equation: in this dynamic, the threat of churn was more powerful than the threat of a competitive bid. They were more afraid of losing us than of being underpriced.

The new analytics module reframed itself in my mind as a tradeable asset. They wanted us to adopt it. We could adopt it in exchange for commercial concessions—a deal structure that gave us additional capability while giving them the adoption metric they needed for their fiscal year story.

The stock price pressure created urgency on their side that I couldn't have manufactured otherwise. The leadership team needed to show revenue stability to a market that had just punished a disappointing earnings report. That made a multi-year commitment from us substantially more valuable than it might normally be. They would pay for the certainty in commercial terms because their internal narrative required it.

And the improved alternatives meant our BATNA was better than the supplier probably realized. They were operating on an outdated assessment of our switching costs. We weren't operating on the same outdated assessment, which gave us confidence to push harder than we might have otherwise.

The negotiation strategy that emerged from synthesizing all of this was: propose a three-year renewal (addressing their retention and revenue stability concerns) with the new analytics module included at no additional cost (giving them the adoption metric they needed) in exchange for a meaningful price reduction and improved service levels. This was a deal structure that created real value for both parties. We got better economics and an additional capability, they got a long-term commitment and a reference customer for their new module.

Without the AI-powered intelligence, I would not have identified most of these leverage points. I certainly wouldn't have known about the CEO's explicit retention priority, the stock price pressure, or the specific timing of the analytics module launch. Each of those insights came from AI-synthesized analysis of publicly available information that I simply wouldn't have had time to read and synthesize manually.

The intelligence didn't negotiate the deal. I did.

The preparation didn't replace the judgment required to read the room, manage the emotional dynamics, and find the creative solution. That was human work. The intelligence gave me the insights that made the creativity possible, the confidence that made the positions credible, and the understanding of the supplier's situation that made the relationship stronger rather than weaker as a result of the negotiation.

THE HUMAN INTELLIGENCE LAYER: WHAT AI CAN'T GIVE YOU

I've been singing the praises of AI-powered market intelligence for several thousand words now, so let me balance the ledger with something important:

There is an entire category of market intelligence that AI cannot provide, and it is often the most valuable intelligence in the dossier.

I'm talking about human intelligence—the insights that come from conversations with people who actually operate in the market you're researching. The supplier account manager who, over coffee at an industry conference, mentions that their company just lost a major account and is under pressure to fill the capacity gap. The procurement colleague at another company who tells you what they're actually paying for the same category you're sourcing. The industry consultant who shares a perspective on where the market is heading that won't show up in any analyst report for another six months. The supplier's former employee who, now that they've moved on, is willing to give you an honest assessment of the company's operational challenges.

This human intelligence layer is irreplaceable. AI can tell you what's in the public record. It can synthesize published information at superhuman speed. But it cannot replicate the trust-based information exchange that happens between professionals who know each other, respect each other, and are willing to share candid perspectives that they'd never put in writing.

The smart way to integrate AI and human intelligence is sequential. Use AI to build your eighty-percent foundation—the market map, the financial analysis, the competitive landscape, the cost driver assessment. Then use that foundation to guide targeted human intelligence gathering that fills the critical twenty percent.

The AI tells you what questions to ask.
The human conversations give you the answers that no algorithm can access.

Here's a practical example. I was preparing for a sourcing event in a specialized technology category. The AI-powered market analysis identified seven potential suppliers, provided financial profiles and capability assessments for each, and generated a competitive landscape map. But one of the seven, a mid-sized company that looked strong on paper, had a reputation question mark. Multiple sources mentioned the company in generally positive terms, but there was a conspicuous absence of reference customers in our industry. The AI could tell me what was publicly known. It couldn't tell me why this supplier hadn't penetrated our industry despite appearing to have the capability.

Two phone calls answered the question. A procurement director at a company in our industry who had evaluated this supplier eighteen months earlier told me that they had failed to deliver on their implementation commitments—not dramatically, but enough that the reference customer who was supposed to be their industry showcase had declined to serve as a reference. An industry consultant confirmed that the supplier was struggling with the regulatory requirements specific to our sector and was six to twelve months away from being ready for our market.

That intelligence, which took about thirty minutes to gather and which no AI tool could have provided, completely changed my evaluation of this supplier. They went from a contender to a "revisit in a year" note in my file. And the two conversations were guided directly by the AI's analysis, which had identified the gap in the first place.

The sourcing professionals who are most effective at market intelligence don't choose between AI and human research. They use both, in sequence, with each informing the other. The AI accelerates the broad research and identifies the gaps. The human conversations fill the gaps with the kind of trust-based, candid, contextually rich intelligence that algorithms can't access.

BEYOND THE NEGOTIATION: INTELLIGENCE FOR CATEGORY STRATEGY

Everything we've discussed so far in this chapter has been framed around negotiation preparation—building intelligence to support a specific commercial conversation. But the same capabilities and workflows apply to a broader, more strategic use case: category strategy development.

When you have AI-powered market intelligence capability, you can do something that most procurement organizations have aspired to but few have achieved consistently: develop market-informed category strategies for every significant spend category, not just the top three or four that get enough attention to justify a full research effort.

Most procurement organizations have a tiered category strategy approach. Their top ten or fifteen categories get formal, comprehensive sourcing strategies that are reviewed annually and supported by market intelligence.

The next twenty or thirty categories get a lighter version—maybe a basic strategy refresh every couple of years, with limited market research. And the long tail (sixty or seventy percent of categories by count, even if not by spend) gets essentially no strategic attention. The reason is bandwidth. Manual market research and category strategy development are time-intensive enough that organizations have to triage. The categories below the cutoff get triaged out.

AI changes the economics of category strategy development. When you can build a comprehensive market assessment in two hours instead of two weeks, suddenly it's feasible to develop informed strategies for categories that previously flew below the radar. That $800,000 translation services category that nobody has looked at in three years? You can have a market assessment and preliminary sourcing strategy by tomorrow afternoon. The $1.2 million facilities maintenance contract that's been on auto-renew for the past two cycles? A competitive landscape analysis and alternatives assessment will take you about ninety minutes.

I worked with a procurement team that used AI-powered market intelligence to systematically assess every category in their portfolio above $500,000 in annual spend —forty-seven categories in total. Using the traditional research approach, this would have been a six-to-nine month project. Using AI-augmented research, supplemented by targeted human intelligence for the most strategic categories, they completed the assessment in about six weeks. The result was a prioritized sourcing roadmap that identified approximately $14 million in addressable savings opportunities across categories that the organization hadn't strategically sourced in years—in some cases, ever.

The AI didn't identify those savings. It identified the market conditions that made the savings possible. The procurement team identified the savings by applying their expertise to the AI's market intelligence and developing strategies the intelligence supported.

THE COMPETITIVE INTELLIGENCE ETHICS QUESTION

Before we close this chapter with the intelligence practice argument, I want to address something that sits at the intersection of market intelligence and professional ethics: where is the line between legitimate competitive intelligence and practices that cross into ethically questionable territory?

This question becomes more pressing as AI makes intelligence gathering easier and more comprehensive. When you can synthesize a supplier's entire digital footprint in an hour—their financial filings, their employees' LinkedIn profiles, their patent applications, their job postings, their social media presence, their leadership's public speaking engagements—the amount of intelligence you can build about a counterparty is unprecedented. And the question of what's appropriate to gather, and how you use what you gather, deserves thoughtful attention.

My framework is straightforward. Anything that is publicly available (financial filings, published interviews, analyst reports, press releases, publicly visible LinkedIn profiles, patent databases, regulatory filings) is fair game.

This is information that the supplier has made available to the public, and analyzing it is no different from what any competent researcher would do.

Where I draw the line is at deception and confidentiality violations. Using AI or any other tool to access information that is not publicly available—proprietary data, confidential internal documents, communications that were not intended for external audiences—is a violation of professional ethics regardless of how easy the technology makes it. Similarly, using information that was shared with you in confidence by a supplier—their cost structure, their pricing methodology, their strategic plans—for the benefit of a competing supplier is an ethical violation that we discussed in Chapter Two's ethics section, and it remains a violation whether you do it manually or with AI assistance.

The ease of intelligence gathering that AI provides doesn't change the ethical principles. It just makes it more important that the best buyers have clear, internalized standards for what's appropriate—because the technology will let you do things that professional ethics should prevent you from doing.

GATHER BROADLY. ANALYZE DEEPLY. USE WISELY

With those principles in place, the intelligence capability becomes a structural advantage rather than a liability. And that's where the chapter is heading next—not just the dossier built before a single negotiation, but the practice maintained continuously across every supplier relationship and every category that matters to your organization.

BUILDING AN INTELLIGENCE PRACTICE, NOT JUST A ONE-TIME EXERCISE

There's an important distinction between doing AI-powered market intelligence and building an AI-powered market intelligence practice. The one-time dossier is valuable. The ongoing intelligence practice is transformative.

The most sophisticated procurement organizations I work with are not just building supplier dossiers before negotiations. They're building AI-powered systems and workflows that track their strategic suppliers and key categories on an ongoing basis, surfacing relevant changes as they happen rather than waiting until a contract renewal forces a research sprint.

What does this kind of continuous intelligence monitoring look like in practice?

It looks like a weekly AI-generated briefing on your top twenty suppliers, flagging any earnings results, leadership changes, major contract wins or losses, regulatory actions, or analyst rating changes that occurred in the previous week. It looks like quarterly category intelligence updates that track pricing trends, competitive landscape shifts, and emerging alternatives in your most strategic spend categories. It looks like automated alerts when a supplier in your portfolio experiences a financially significant event, like a credit downgrade, a major litigation, or a CEO departure that might affect their ability to perform on your contract.

None of this existed five years ago. Most of it wasn't even conceptually feasible. Today, a single procurement professional with the right AI tools and the right workflow can maintain a level of market and supplier intelligence that previously required a dedicated research team.

This is the democratization of intelligence that I described in Part One. The individual sourcing professional who used to be at an information disadvantage against a well-resourced sales team can now walk into the room comparably informed. The mid-sized company that couldn't afford a market intelligence subscription can now generate comparable analysis using AI tools that cost a fraction of the price. The category manager who used to rely on gut feel and outdated benchmarks can now ground every conversation in current, evidence-based market intelligence.

The procurement professionals who build this capability into their routine—who treat market intelligence not as a pre-negotiation sprint but as a continuous practice—are developing a competitive advantage that compounds over time. Every intelligence cycle makes them more knowledgeable about their markets, more attuned to the signals that matter, and more effective at translating information into outcomes. They are, to borrow the language from Chapter Two, becoming more dangerous in negotiations because they're getting smarter. Continuously, systematically, and at a pace that their less informed competitors simply cannot match.

The information asymmetry that used to favor well-resourced sellers is flipping. The procurement professionals who are investing in the skills and tools to build continuous market intelligence are going to be the ones who consistently deliver superior outcomes for their organizations.

Know more than the other side of the table. It's always been the most reliable path to better procurement outcomes. AI just made it more accessible.

Now let's talk about a dimension of intelligence that keeps procurement leaders awake at night—the risks hiding in your supply chain that you can't see yet.

Let's talk about supplier risk in real time.

CHAPTER 6
SUPPLIER RISK IN REAL TIME

FROM HOPING FOR THE BEST TO SEEING AROUND CORNERS

On March 11, 2011, a magnitude 9.0 earthquake struck off the coast of Japan, triggering a tsunami that devastated the country's northeastern coastline and caused one of the worst nuclear disasters in history at the Fukushima Daiichi power plant. The human toll was catastrophic—nearly twenty thousand people lost their lives. But the earthquake also triggered a supply chain catastrophe that rippled across the globe for months, affecting industries that most people would never have connected to a natural disaster in Japan.

Here's what most executives in the automotive and electronics industries learned in the weeks following that earthquake: they had no idea how dependent they were on suppliers in the affected region. Not their direct suppliers—most companies knew where their tier-one suppliers were located. But their suppliers' suppliers. The small, specialized component manufacturers three and four tiers deep in the supply chain that nobody in procurement had ever heard of, let alone mapped, let alone assessed for geographic concentration risk.

A single company, Renesas Electronics, manufactured approximately forty percent of the microcontrollers used in automobiles worldwide. Their primary fabrication facility was in the earthquake zone. When it went offline, automotive production lines around the world started shutting down in a slow cascade over the following weeks as buffer stocks were depleted and no replacement supply could be found. General Motors, Toyota, Ford, Honda, companies with some of the most sophisticated supply chain operations on the planet, found themselves unable to build cars because a component they'd never thought much about, manufactured by a company most of their procurement teams had never directly engaged with, was suddenly unavailable.

The semiconductor shortage that followed the earthquake lasted months. A decade later, the same kind of concentrated vulnerability surfaced again at much larger scale: AlixPartners estimated that the 2021 semiconductor shortage cost the global automotive industry $210 billion in lost revenue in that year alone. The root cause in both cases was not a failure of procurement execution. It was a failure of risk visibility. These companies didn't know what they didn't know. They couldn't see the risk because they'd never mapped the supply chain deeply enough to identify where the concentration existed.

I tell this story at the beginning of this chapter because

it illustrates something fundamental about supply chain risk that most organizations still haven't fully internalized, even after the earthquake, even after COVID, even after the semiconductor crisis, even after the Suez Canal blockage, even after the series of supply chain disruptions that have made "supply chain" a topic of dinner party conversation for the first time in history: the risks that hurt you most are the ones you don't see coming.

And that (seeing the risks before they materialize, monitoring the signals before they become crises, shifting from reactive damage control to proactive risk management) is where artificial intelligence is creating one of its most profound and practically valuable transformations in procurement.

A note for senior leaders before we go further:

The argument of this chapter is not that AI makes risk monitoring incrementally better. It's that the historical economics of supplier risk management (periodic assessments, expensive consulting engagements, manual data gathering that quickly went stale) kept rigorous risk monitoring out of reach for most organizations and most of the supply base. AI changes the economics. Continuous monitoring of three thousand suppliers across financial, geographic, cybersecurity, quality, and regulatory dimensions is now economically feasible at a scale that was impossible five years ago.

The strategic question for procurement leadership is whether your organization will treat supplier risk management as a structural insurance investment that prevents catastrophic loss, or continue to treat it as a periodic compliance exercise. The chapter that follows is the playbook for the first option.

THE RISK MANAGEMENT PROBLEM: WHY TRADITIONAL APPROACHES AREN'T ENOUGH

Let's be honest about what supplier risk management has actually looked like in most organizations for most of the profession's history. It has looked like a spreadsheet.

I'm not being dismissive. I've built those spreadsheets. I've maintained those spreadsheets. I've presented those spreadsheets to executive leadership teams and used them to drive risk mitigation decisions. (If you've ever sat in a meeting where someone presents a sixty-tab supplier risk workbook and asks you to take it home and review the methodology, you have my deepest professional sympathy. I have been on both sides of that meeting and neither side is good.) Supplier risk assessment matrices—typically organized by risk category (financial, operational, geographic, quality, cybersecurity, regulatory) with probability and impact ratings—have been a staple of procurement practice for decades, and they have real value. They force organizations to think systematically about risk. They create a framework for prioritization. They provide a basis for conversation between procurement and the business about which supplier risks warrant investment in mitigation.

But they have a fundamental limitation that no amount of formatting or analytical sophistication can overcome:

they are snapshots. A supplier risk assessment conducted in January tells you what the risk landscape looked like in January. It tells you nothing about what changed in February. The supplier who was financially stable in January might issue a profit warning in March. The region that was politically stable in January might experience a regime change in April. The factory that passed its quality audit in January might have a catastrophic fire in May.

Most organizations conduct formal supplier risk assessments annually. Some do it semi-annually. A few do it quarterly. But even quarterly assessment leaves ninety-day gaps during which the risk landscape can change significantly without anyone in procurement knowing about it. And in practice, the assessment process itself takes weeks of data gathering, supplier questionnaires, financial analysis, and review. By the time the assessment is complete, the information is already aging.

The result is a risk management approach that I've come to describe as "hoping for the best between assessments." You do the rigorous work periodically. You identify and mitigate the risks you can see. And then you go back to your sourcing and negotiation work and hope that nothing changes significantly before the next assessment cycle.

When something does change (when a supplier files for bankruptcy, when a critical component becomes unavailable, when a geopolitical event disrupts a supply route) you discover it when it hits your operations, not when the early warning signals first appeared. By that point, your options are limited, your costs are high, and your organization is in reactive crisis mode.

This isn't a criticism of the procurement professionals

who've been doing this work. It's a criticism of the tools they've had available. You can't monitor a thousand suppliers continuously with spreadsheets and annual questionnaires. The math doesn't work. The human bandwidth doesn't exist. The data isn't available in real time.

Imagine Big Ron with this kind of risk visibility.

He didn't have it, of course. Big Ron's risk monitoring was a sixty-tab spreadsheet, a quarterly check-in call to his top fifteen suppliers, and the certainty that any real problem would announce itself when it hit his shipping dock. His instincts were sharp—twenty-five years of watching suppliers fail teaches you what failure looks like —but he had the bandwidth to monitor maybe fifteen relationships closely. The rest of the portfolio was a black box that surfaced a crisis from time to time and got handled, badly, with whatever leverage he could improvise in the moment.

Give Big Ron the kind of visibility this chapter describes—financial health signals across his suppliers, geopolitical events flagged against his sub-tier map, concentration risks he could actually see—and he would have been transformed. Not as a person. Big Ron wasn't going to soften because his tools got better. As an operator, though, he would have been faster, sharper, and harder to surprise. A Big Ron with four months of advance warning on a supplier's financial distress could have helped that supplier through it, or extracted concessions for the privilege of his continued business. A Big Ron who could see concentration risk forming across his portfolio could have built resilience, or used the leverage. Which one he chose was always going to be up to him. Same as it'll be up to you.

Until now.

There's a Minority Report quality to good supplier risk monitoring that's worth naming. Tom Cruise's precrime unit didn't catch criminals after the fact; they saw the murder coming and intervened before it happened. The whole point of the movie, of course, is that the precogs weren't infallible—sometimes they were wrong, sometimes the system got manipulated, and the cost of acting on a prediction that turned out to be false was real. Continuous supplier risk monitoring sits in roughly the same territory. The system surfaces signals that a supplier is heading for trouble. Some of those signals will be right and the intervention will save you from a disruption that hadn't happened yet. Some of them will be wrong, and acting on a false signal—pulling business from a supplier who was actually fine, putting an alternative on standby that wasn't needed—costs you something. The job isn't to act on every signal. It's to have the signals at all, and to develop the judgment that separates the ones worth acting on from the ones worth watching.

THE AI TRANSFORMATION: CONTINUOUS RISK MONITORING

AI transforms supplier risk management from a periodic assessment exercise into a continuous monitoring capability. The difference is not incremental. It's qualitative. It's the difference between getting an annual physical and wearing a medical monitor that tracks your vital signs every second of every day and alerts you the moment something moves outside normal parameters.

Let me walk you through the capabilities that make this transformation possible.

Financial Risk Monitoring

Supplier financial distress is one of the most predictable and most consequential risks in procurement. It's also one of the areas where AI monitoring is most mature. AI-powered financial risk monitoring systems continuously analyze a supplier's financial health indicators (revenue trends, margin trajectory, debt ratios, cash flow patterns, credit ratings, payment behavior with their own suppliers) and generate risk scores that update in real time as new information becomes available.

These systems can detect deterioration months before it becomes obvious through traditional channels. A supplier whose payment behavior with their own suppliers starts slowing. A supplier whose credit rating gets a quiet downgrade. A supplier whose accounts receivable are aging in ways that suggest collection problems. A supplier whose hiring patterns suggest cost cutting in operationally critical areas. Each of these signals individually might be inconclusive. Tracked continuously and analyzed in combination, they can paint a picture of financial stress that gives procurement teams months of warning before a supplier failure actually occurs.

I worked with a CPO who described how their AI-powered financial risk monitoring system had flagged a critical supplier's deteriorating financial signals approximately four months before that supplier publicly disclosed financial difficulties. By the time the public disclosure came, the procurement team had already qualified an alternative source, negotiated standby capacity, and built a contingency plan for managing the transition if it became necessary. When the supplier ultimately filed for bankruptcy reorganization six months later, the CPO's organization was the only major customer that experienced no operational disruption—because they'd been preparing for this scenario for half a year while their competitors were still receiving reassuring quarterly business reviews.

Geopolitical and Geographic Risk Monitoring

Geopolitical risk has become one of the most challenging dimensions of supplier risk management, and one of the most difficult to monitor through traditional means. Recent trade disputes, sanctions, regional conflicts, port closures, and infrastructure attacks have demonstrated the need for continuous monitoring of geopolitical conditions in every region where critical suppliers operate.

AI-powered geopolitical monitoring tools can scan global news sources, government announcements, regulatory filings, social media, and specialized intelligence feeds to identify events that might affect specific suppliers or supply routes. They can correlate these events to your specific supply chain map, automatically flagging that a port closure in a specific country affects three of your suppliers, that a trade policy change creates new tariff exposure for a specific category, or that political instability in a specific region threatens your tier-two supply for a critical component.

Resilinc, one of the leading platforms in this space, monitors millions of news sources globally and uses natural language processing to identify supply-chain-relevant events as they occur—sometimes within minutes of the first news reports. Their EventWatch capability has flagged everything from natural disasters to factory fires to regulatory actions, providing customers with hours or days of advance warning before the same events would have surfaced through traditional intelligence channels.

Cybersecurity Risk Monitoring

Cybersecurity risk has become one of the fastest-growing dimensions of supplier risk, and AI-powered monitoring is increasingly essential for managing it. Every supplier who connects to your systems, processes your data, or operates critical infrastructure that affects your operations represents a potential cybersecurity risk—and the threat landscape evolves faster than any periodic assessment can track.

AI-powered cybersecurity risk monitoring platforms like SecurityScorecard, BitSight, and RiskRecon continuously assess suppliers' security postures by analyzing their externally observable security practices, their patching cadence, their network configurations, and their presence in threat intelligence feeds. They can detect when a supplier has been compromised sometimes before the supplier themselves is aware of it, and provide early warning that allows procurement and security teams to take protective action.

The practical implication for procurement professionals is significant. When you're evaluating suppliers (particularly technology suppliers who will have access to your systems and data) AI-powered cybersecurity risk assessment provides a continuously updated view of the supplier's security posture that supplements (and in some cases replaces) the point-in-time security questionnaires that have been the traditional approach. And when you're managing ongoing supplier relationships, continuous cybersecurity monitoring provides early warning of security degradation that should trigger conversation, escalation, or in extreme cases, contingency planning.

Quality and Operational Risk Monitoring

Quality risk—the risk that a supplier's output degrades below acceptable standards—has traditionally been managed through periodic audits, incoming quality inspection, and performance tracking against contractual metrics. These approaches catch quality problems after they occur. AI-powered quality risk monitoring aims to detect them before they materialize.

How? By analyzing patterns in the data that precede quality failures. Machine learning models trained on historical quality data can identify the leading indicators of quality degradation (subtle shifts in manufacturing parameters, changes in raw material sourcing, increases in rework rates, staffing changes in quality-critical roles) that human reviewers might not notice until the defect rate has already climbed to unacceptable levels.

This is particularly valuable in categories where quality failures have severe consequences. In pharmaceutical manufacturing, a quality failure can result in a product recall that costs hundreds of millions of dollars and endangers patient safety. In automotive manufacturing, a quality failure can result in the kind of catastrophic recall we discussed in the introduction with the Takata airbag scandal. In food and beverage, a quality failure can result in illness, death, and brand destruction.

For procurement professionals in these industries, the ability to detect quality risk signals before they become quality failures is not just a nice-to-have capability. It is a fundamental shift in how supplier quality is managed—from catching problems and cleaning up the aftermath to identifying risk patterns and intervening proactively.

Regulatory and ESG Risk Monitoring

There's a dimension of supplier risk that has exploded in importance over the past five years and that AI is uniquely suited to monitor: regulatory and Environmental, Social, and Governance risk.

The regulatory landscape governing supply chains has become significantly more complex in recent years. Legislation like the German Supply Chain Due Diligence Act, the EU Corporate Sustainability Due Diligence Directive, the Uyghur Forced Labor Prevention Act in the United States, and the proliferating regulations around carbon disclosure and emissions reporting have created obligations that extend deep into supplier networks. Organizations are now legally and reputationally responsible for practices occurring not just in their direct suppliers' operations, but in their suppliers' suppliers' operations, including labor practices, environmental impacts, human rights, and governance standards across extended supply chains that span dozens of countries and thousands of entities.

Monitoring compliance with these obligations manually is essentially impossible at scale. The regulations are complex, they vary by jurisdiction, they change frequently, and they require continuous monitoring rather than point-in-time assessment. AI-powered regulatory risk monitoring addresses this challenge by continuously scanning regulatory databases, government enforcement actions, NGO reports, media coverage, and supplier disclosures to identify potential compliance issues across the supply base.

A consulting engagement I worked on involved an organization using AI-powered ESG monitoring when the system flagged a tier-two supplier in Southeast Asia that had appeared in an NGO report alleging labor practice violations. The organization's tier-one supplier, the one they had a direct relationship with, had not disclosed this issue and may not have been aware of it. The AI-powered monitoring system identified it through natural language processing of an NGO report published in a language that nobody on the procurement team could read, cross-referenced it against the supply chain map, and generated an alert that triggered an investigation.

The investigation confirmed the concern. The organization worked with their tier-one supplier to address the issue—which ultimately resulted in the tier-two supplier implementing improved labor practices under a corrective action plan. The alternative (discovering the issue six months or a year later, through a journalist's investigation or a regulatory inquiry) would have meant reputational damage, potential legal liability, and a much more costly remediation process.

This is an area where the value of AI-powered monitoring extends well beyond operational risk into brand protection, legal compliance, and corporate responsibility. The organizations that can demonstrate continuous, AI-powered monitoring of their supply chains for ESG risks are not only managing regulatory compliance more effectively—they're building credibility with investors, customers, and regulators who are increasingly demanding evidence of proactive supply chain governance.

Concentration Risk Analysis

Here's a type of risk analysis that AI makes possible at a scale that manual approaches never could: portfolio-level concentration risk assessment. This is the practice of analyzing your entire supply base (not supplier by supplier, but as a system) to identify patterns of concentration that create vulnerability.

Traditional risk assessment looks at suppliers individually: is this supplier financially healthy? Is that supplier geographically vulnerable? AI-powered concentration risk analysis looks at the aggregate picture: how many of your critical components are sourced from a single geographic region? How many of your strategic suppliers are dependent on the same raw material provider? How many of your technology platforms are built on the same underlying infrastructure? If a single event (a natural disaster, a trade policy change, a technology platform failure) disrupted a specific node, how many of your supply chains would be affected simultaneously?

These are questions that are nearly impossible to answer at scale without AI, because they require analyzing relationships and dependencies across thousands of suppliers, hundreds of categories, and dozens of geographic and technological dimensions simultaneously. No human team can hold that level of complexity in their heads. But an AI system can map the dependencies, model the scenarios, and identify the concentration risks that create systemic vulnerability.

I saw this capability demonstrated at a conference where a CPO described how AI-powered concentration risk analysis had revealed that three of their "independent" suppliers for a critical electronic component were all sourcing from the same sub-tier manufacturer in Taiwan. On paper, the organization had a well-diversified supply base for this component—three qualified suppliers, each under contract, each delivering reliably. In reality, they had a single point of failure that none of them had identified through traditional supplier-by-supplier risk assessment. The AI identified it by mapping the sub-tier supply chain connections across all three suppliers simultaneously and flagging the common dependency.

They diversified their sub-tier exposure before it became a problem. Their competitors, who had the same concentration risk and didn't know about it, discovered it during the semiconductor shortage when all three of their "independent" suppliers experienced the same disruption at the same time—because they all depended on the same source.

This pattern is more common than most procurement organizations realize, and the consequences are real. A consulting engagement I worked on uncovered, through AI-powered mapping, that two of three qualified suppliers for a critical input material were both sourcing their base component from the same supplier facility in Louisiana—a facility that happened to sit in a flood zone. When Hurricane Ida hit in 2021, that facility was damaged and both suppliers lost their input supply simultaneously. The organization's third supplier, the one that happened to use a different upstream source, couldn't scale fast enough to cover the gap. The resulting supply disruption cost approximately $3 million in expedited sourcing and production delays.

After the disruption, the organization implemented sub-tier mapping that would have revealed the shared dependency before the hurricane. The $3 million lesson was that "three suppliers" is not the same as "three independent supply chains," and only AI-powered mapping can reliably tell you the difference.

A practical warning before we move on. The noise-to-signal problem is going to be your main challenge with all of this. Every one of the capabilities I just walked through will produce alerts. Most of those alerts will be wrong, irrelevant, or both. Getting good at the practice means getting good at ignoring most of what shows up on the dashboard—and, over time, recognizing which of the things you'd normally ignore is the one that matters.

SCENARIO MODELING: PREPARING FOR WHAT HASN'T HAPPENED YET

One of the most sophisticated applications of AI in supply chain risk management is scenario modeling—using AI to simulate potential disruption scenarios and model their impact on your supply chain before they actually occur.

Traditional scenario planning is a valuable but labor-intensive exercise. You gather a cross-functional team, spend a day or two brainstorming potential disruption scenarios, manually estimate the impact of each one, and develop contingency plans for the most significant risks. It's useful work, but it's limited by the team's imagination, constrained by the scenarios they think to model, and typically conducted once a year at most.

AI-powered scenario modeling can run thousands of simulations continuously, testing the resilience of your supply chain against a vast range of potential disruptions—natural disasters in every region where you have suppliers, financial failures at every critical supplier, trade policy changes affecting every relevant trade route, commodity price spikes in every relevant market. The models can estimate the cascading effects of disruptions across your supply chain, identifying the bottlenecks and single points of failure where a disruption at one node would cascade into widespread operational impact.

The practical value of this capability is in the preparation it enables. When the model shows you that a natural disaster in a specific region would disrupt supply for forty-seven percent of your electronic components, that's a signal to invest in geographic diversification, build safety stock, or qualify alternative sources in other regions—now, before the disaster happens, while you have time and options. When the model shows you that the bankruptcy of a specific supplier would cascade through your supply chain and affect production across three business units, that's a signal to monitor that supplier's financial health more closely, develop a contingency plan, and potentially accelerate the qualification of a backup source.

Scenario modeling doesn't predict the future. It illuminates vulnerability. And illuminated vulnerability is the starting point for intelligent, proactive risk mitigation—the kind that prevents crises rather than responding to them.

BUILDING A RISK INTELLIGENCE PRACTICE: THE PRACTICAL FRAMEWORK

Let me move from capabilities to practice, because understanding what AI can do for risk monitoring is less useful than understanding how to actually implement it in your organization. The framework I've developed and refined over several years has four layers, and each layer builds on the one below it.

Layer One: Know What You're Monitoring

This sounds obvious, but it's where most organizations stumble out of the gate. Before you can monitor supplier risk effectively, you need to know which suppliers matter enough to monitor—and "all of them" is not a useful answer. A typical large organization might have three thousand to five thousand active suppliers. Monitoring all of them continuously with the same level of attention would overwhelm any team and any system, generating so many alerts that the important signals would drown in noise.

The segmentation framework from Chapter One is your starting point. Your strategic suppliers in the upper right quadrant of the criticality-versus-spend matrix get the highest level of monitoring intensity. These are the suppliers whose failure would directly and immediately impact your operations, and they deserve continuous, multi-dimensional monitoring across financial, operational, cybersecurity, and geographic risk dimensions.

Your next tier (high-spend but lower-criticality suppliers, and lower-spend but high-criticality suppliers) get moderate monitoring. Financial health tracking, basic news monitoring, and geographic risk mapping, but not the same depth of continuous assessment that your strategic suppliers receive.

Your tail suppliers (the long tail of low-spend, low-criticality vendors) get minimal proactive monitoring. This isn't because risk doesn't exist in the tail. It does. But the economics of monitoring thousands of small suppliers continuously don't justify the investment. For the tail, automated financial health scoring and exception-based alerting (notify me if something significantly changes) is usually sufficient.

The key decision here is segmentation, and getting it right matters. I've seen organizations waste enormous amounts of risk management bandwidth monitoring suppliers that didn't warrant it, while neglecting critical suppliers because they were buried in the general population. Take the time to segment properly before you build the monitoring program.

Layer Two: Define What You're Monitoring For

Risk monitoring that doesn't translate into actionable intelligence is expensive noise. For each supplier tier, you need to define the specific risk indicators you're tracking and the thresholds that trigger action.

For a strategic technology supplier, your monitoring dashboard might include:

1. Credit rating and financial health score (trigger: any downgrade or score decline exceeding ten percent),
2. Stock price trajectory (trigger: decline exceeding fifteen percent in thirty days),
3. Leadership stability (trigger: departure of CEO, CTO, or your account executive),

4. Cybersecurity posture score (trigger: any decline below your minimum threshold),
5. News sentiment (trigger: significant negative coverage from credible sources),
6. Regulatory actions (trigger: any material regulatory action or investigation),
7. Geographic risk events (trigger: any natural disaster, political instability, or infrastructure disruption within fifty miles of a critical facility), and
8. Customer satisfaction signals (trigger: significant negative reviews or public customer complaints).

Each trigger should have a defined response protocol. A credit downgrade for a strategic supplier doesn't just generate an alert—it triggers a review meeting with the category manager, an assessment of backup options, and potentially a conversation with the supplier about their financial situation and continuity planning. A cybersecurity score decline triggers an immediate assessment of what data and system access the supplier has and whether enhanced monitoring or access restrictions are warranted.

The discipline of defining indicators and thresholds in advance, rather than trying to interpret signals in the moment, is what separates proactive risk management from reactive panic. It's the difference between a fire department that has evacuation plans and sprinkler systems and one that waits until the building is burning to figure out where the exits are.

Layer Three: Build the Information Architecture

This is the most technically demanding layer, and it's the one where purpose-built procurement risk platforms add the most value compared to general-purpose AI tools.

Continuous risk monitoring requires connecting multiple data sources into a unified monitoring framework, including:

- Financial data providers.
- News aggregation services.
- Government and regulatory databases.
- Shipping and logistics data.
- Weather and natural disaster monitoring services.
- Cybersecurity rating platforms.
- Social media sentiment analysis.
- Supplier-provided data from performance management systems and supplier portals.
- Internal data from your own procurement, quality, and financial systems.

The AI layer sits on top of this data infrastructure, continuously analyzing inputs from all sources, correlating signals across dimensions, and generating risk scores and alerts that reflect the integrated picture rather than siloed individual indicators.

If your organization has the budget and the strategic commitment, purpose-built supply chain risk platforms like Resilinc, Everstream Analytics, or the risk modules within major source-to-pay suites like Coupa and SAP Ariba, all provide pre-built integrations with these data sources and pre-configured risk models that can be deployed relatively quickly. These platforms represent the state of the art in supply chain risk monitoring, and for organizations with complex, global supply chains and significant supply chain risk exposure, they can deliver substantial value.

If your organization doesn't have the budget for a dedicated risk platform, or if you're in the early stages of building the case for one, you can build a meaningful risk monitoring capability using general-purpose AI tools and publicly available data sources. It won't be as comprehensive or as automated as a purpose-built platform, but it can provide significantly more risk visibility than the annual spreadsheet assessment that most organizations are currently working with.

Here's a practical approach. Set up a weekly workflow where you use a generative AI tool to scan for significant developments related to your top twenty strategic suppliers. Feed the AI a list of your suppliers and ask it to identify any significant news, financial developments, leadership changes, regulatory actions, or market events from the past week that could affect their ability to serve you. Review the output, assess the significance of anything flagged, and take action on the items that warrant it.

This isn't enterprise-grade risk monitoring. It's a procurement professional with a laptop and a subscription spending ninety minutes a week to maintain a level of supply chain awareness that they've never had before.

Let's be real... It's imperfect. It's incomplete.

But it's significantly better than what most organizations have today.

Layer Four: Connect Risk Intelligence to Decision-Making

This is where the rubber meets the road—and where many risk management programs stumble. Having great risk intelligence is useless if it doesn't change decisions. And changing decisions requires that risk intelligence be integrated into the procurement processes where decisions are actually made: sourcing strategy, supplier selection, contract negotiation, and ongoing supplier management.

In practice, this means that risk intelligence should be an input to every significant procurement decision, not a parallel process that runs alongside procurement but doesn't actually influence it.

When you're developing a sourcing strategy, the risk assessment for the category should be on the table alongside the spend analysis and market intelligence. When you're evaluating supplier responses to an RFP, the risk profile of each supplier should be a weighted factor in the evaluation—not just their price, capability, and references. When you're negotiating a contract, the risk intelligence should inform the contractual protections you prioritize—business continuity provisions, financial health reporting requirements, cyber security standards, and termination rights linked to risk triggers.

I've worked with organizations that had excellent risk intelligence and terrible risk outcomes—because the intelligence sat in a dashboard that nobody integrated into their decision-making processes. The risk team produced beautiful reports. The sourcing team made decisions without reading them. The contract team negotiated terms without considering the risk profile. And when a supplier failed, everyone expressed surprise and pointed fingers, despite the fact that the warning signs had been documented and ignored.

The integration of risk intelligence into procurement decision-making is not a technology problem. It's an organizational design problem. It requires that risk considerations be embedded in sourcing templates, evaluation frameworks, contract checklists, and governance processes—not as optional inputs but as mandatory checkpoints that procurement professionals must acknowledge and address before proceeding.

THE SUB-TIER VISIBILITY CHALLENGE

The Renesas story from the beginning of this chapter illustrates the challenge of sub-tier supply chain visibility. Even the best risk monitoring program is limited if it only covers your direct, tier-one suppliers. And the risks that cause the most catastrophic disruptions often originate deeper in the supply chain—at the tier-two, tier-three, and tier-four levels that most procurement organizations have never mapped.

Sub-tier visibility has been one of the most persistent challenges in supply chain management for decades. The fundamental difficulty is straightforward: your tier-one suppliers are often reluctant to share detailed information about their own supply chains because they view that information as proprietary and competitively sensitive. Even when they're willing to share, they may not have comprehensive visibility into their own supply chains for the same reasons you don't have visibility into yours.

AI is beginning to crack this problem, though we're still in the early stages.

There are several approaches that are showing promise.

First, AI-powered supply chain mapping tools can build approximate sub-tier supply chain maps by analyzing publicly available data—import/export records, corporate ownership databases, product composition data, patent filings, regulatory disclosures, and industry-specific databases. These maps are imperfect, they represent a best estimate rather than a verified reality, but they provide significantly more visibility than the complete blind spot that most organizations have today at the sub-tier level.

Second, collaborative supply chain platforms are creating models where suppliers share supply chain data in a structured, controlled way—revealing enough about their sub-tier supply chains to enable risk mapping without exposing competitively sensitive details. AI plays a role in these platforms by normalizing and analyzing the shared data, identifying concentration risks and single points of failure across the extended supply chain.

Third, some organizations are beginning to use AI-powered analysis of their own procurement data, combined with industry knowledge bases, to infer sub-tier supply chain dependencies. If you know that your tier-one supplier uses a specific manufacturing process that requires a specific rare earth mineral, and you know that eighty percent of that mineral's global supply comes from a single region, you can make a reasonable inference about your sub-tier geographic concentration risk even without explicit supply chain data from your supplier.

None of these approaches provides the complete, verified, real-time sub-tier visibility that supply chain professionals dream about. But each of them provides significantly more visibility than the alternative—which, at most organizations, is none at all.

In supply chain risk management, partial visibility is significantly better than no visibility, because it at least identifies the areas where you need to ask questions, conduct due diligence, and build contingency plans.

THE COVID LESSON: WHAT CONTINUOUS RISK MONITORING CHANGES IN A REAL CRISIS

The COVID-19 pandemic remains the clearest real-world demonstration of the difference between having continuous risk monitoring and not having it during a global supply chain disruption. The pattern has now been studied across hundreds of organizations and validated through multiple post-disruption analyses, and the lesson it taught the procurement profession is worth stating directly.

Organizations that had AI-powered supply chain risk monitoring in place before the pandemic had a fundamentally different experience from those that didn't.

The organizations with monitoring saw the signals early. Resilinc—the platform discussed earlier in this chapter, which uses natural language processing to scan news, social media, and government reports across 160 countries and more than 100 languages—sent its first alerts about an unusual pneumonia outbreak in Wuhan on December 28, 2019. The first supplier confirmation of a disruption arrived on January 4, 2020, the same day Resilinc opened a virtual war room that geofenced Wuhan and identified suppliers in and on the outskirts of the city that sat in their customers' networks. Between January 4 and January 27, when the Chinese government formally locked down Wuhan, Resilinc issued three customer alerts. By the second quarter of 2020, the platform had logged more than 35,000 supplier disruption events globally. Organizations using these systems could map which of their suppliers had operations in the affected regions within hours of an alert—assuming they had taken the time to map their supply chains in advance. They were activating backup suppliers, adjusting inventory positions, and communicating with stakeholders about potential impacts while many of their competitors were still debating whether the situation would affect their operations.

The organizations without monitoring experienced the pandemic the way most organizations did—as a series of unpleasant surprises. A supplier email in late February or early March saying they couldn't fulfill an order because their factory was shut down. A scramble to figure out which other suppliers might be affected. Frantic phone calls to alternative sources that were already overwhelmed with demand from other panicking buyers. Weeks of uncertainty about which supply chains were disrupted, how badly, and for how long.

The difference in outcomes was significant, and McKinsey has the data on it. McKinsey's analysis of pandemic-era supply chain disruption found that a single thirty-day disruption to a vulnerable supply chain produces an EBITDA margin gap of three to five percent. That math gets ugly fast across a portfolio of categories. The organizations that detected disruptions earlier and activated contingency plans faster avoided the worst of the stockouts, kept their inventory positions cleaner, and captured share from competitors who were still figuring out what was happening. McKinsey's annual supply chain leader surveys, run every year since, have tracked how rapidly risk monitoring investment has expanded—though those same surveys keep finding that companies have decent visibility into their tier-one suppliers and almost none into their tier-three suppliers and beyond, which is where most of the painful disruptions actually originate.

To be clear: AI-powered risk monitoring didn't prevent the pandemic from disrupting supply chains. Nothing could have done that. What it did was provide earlier warning, better situational awareness, faster response, and more informed decision-making. It turned a surprise into a recognized challenge—and in crisis management, the difference between being surprised and being prepared, even partially prepared, can determine whether an organization gains or loses market share during a disruption.

THE HUMAN ELEMENT: WHY AI DOESN'T REPLACE RISK JUDGMENT

Throughout this chapter, I've been describing AI-powered capabilities that represent significant improvements over traditional risk management approaches. But I want to close the capabilities discussion with an important caveat that connects directly to the "brilliant intern" model from Chapter Three.

AI is excellent at detecting patterns, monitoring data, and generating alerts. It is not excellent at deciding what to do about them. Risk management decisions (how to respond to a signal, how to balance risk mitigation against cost and relationship considerations, when to escalate and when to monitor, when to invoke a contractual right and when to work collaboratively with a supplier through a difficulty) are judgment calls that require the kind of contextual understanding, relationship awareness, and strategic thinking that AI cannot provide.

A risk monitoring system might flag that a strategic supplier's financial health score has declined. But the appropriate response depends on context that the system doesn't have. Is the decline due to a temporary issue that the supplier is managing, or a structural problem that threatens their viability? Is the supplier transparent about the situation, or are they hiding it? Do you have alternatives that could absorb the volume, or would a disruption be catastrophic? Is the supplier relationship strong enough to survive a conversation about their financial health, or would raising the issue damage a fragile relationship?

These are human questions. They require human judgment. And they require the kind of supplier relationship depth that we discussed in Chapter One—because the quality of the information you get in response to these questions depends entirely on the trust that exists between you and the supplier.

The model that I've seen work at organizations that are leading in supply chain risk management is AI providing the signals and humans providing the judgment. The AI monitors thousands of data points continuously and surfaces the signals that warrant attention. The human reviews those signals, applies organizational context and relationship knowledge, and decides what action is appropriate. The AI handles the surveillance at scale. The human handles the interpretation and response with nuance.

This is the pattern we've been describing throughout Part Two: AI reveals the landscape, human judgment navigates it. In risk management, the landscape is the constellation of threats and vulnerabilities across your supply base. The navigation is the set of decisions about which risks to accept, which to mitigate, and how to prepare for the ones you can't control.

BUILDING YOUR RISK MANAGEMENT CAPABILITY: WHERE TO START

If your organization currently manages supplier risk through annual assessments and spreadsheets (statistically speaking, that's most organizations) the capabilities I've described in this chapter might feel overwhelming. How do you get from where you are, to continuous, AI-powered risk monitoring? The answer, as with most transformation journeys, is one step at a time.

Start with your critical suppliers. As we discussed, don't try to build a comprehensive risk monitoring program for your entire supply base on day one. Identify your ten to twenty most critical suppliers—the ones whose failure would cause the most significant operational impact—and build a monitoring capability for that group first. Even a manually executed weekly scan using generative AI tools, applied to your most critical suppliers, is a meaningful improvement over an annual assessment applied to everyone.

Get your data house in order. Effective risk monitoring requires knowing basic facts about your suppliers—where they're located, what they provide, how dependent you are on them, whether alternatives exist. If your supplier master data is incomplete or inaccurate—and at most organizations, it is—investing in data quality is a prerequisite for effective risk monitoring. You can't monitor risks you can't identify in suppliers you can't see.

Build the business case with a near-miss or an actual disruption. Nothing accelerates investment in risk management like a vivid, costly example of what happens without it. If your organization has experienced a supplier disruption in the recent past, use it as the case study for why proactive monitoring is worth the investment. Walk leadership through the timeline: when did the risk signals first appear, when did the organization become aware of the problem, what was the cost of the lag between those two moments, and how would continuous monitoring have changed the outcome?

Connect risk intelligence to sourcing decisions. Even before you have a sophisticated monitoring platform, you can start integrating risk considerations into your sourcing processes. Add a risk assessment step to your sourcing methodology by requiring an evaluation of the supplier's financial health, geographic concentration, cybersecurity posture, and business continuity capabilities before you award a contract. This doesn't require AI. It requires discipline. And it creates the organizational habit of considering risk alongside cost, quality, and capability in every sourcing decision.

Build toward continuous monitoring. As your risk management maturity grows and the business case strengthens, invest in increasingly automated and continuous monitoring capabilities—whether through purpose-built platforms, enhanced capabilities within your existing procurement technology stack, or AI-augmented manual processes that your team executes on a regular cadence.

The journey from spreadsheet-based annual assessments to continuous AI-powered risk monitoring is a multi-year transformation. But every step along the way delivers value. The organization that scans its top twenty suppliers weekly using AI tools is significantly better positioned than the one conducting annual assessments. The organization that maps its sub-tier supply chain for critical components, even imperfectly, is significantly better positioned than the one that has never looked beyond tier one.

And there's one more dimension of risk management worth naming directly before we close this chapter—a paradox that every risk management professional knows intimately and that shapes how the discipline gets resourced and recognized within an organization.

THE RISK PARADOX: WHY SUCCESS IS INVISIBLE

The better your risk management program works, the less visible its value becomes.

When supply chain risk monitoring detects a signal early, and procurement activates a backup supplier before the disruption materializes, and operations continue without interruption, nobody throws a parade. Nobody writes a case study about the crisis that didn't happen. Nobody presents a savings number to the CFO, because the cost of a disruption that was prevented is inherently unmeasurable. You can estimate it. You can model what would have happened. But you can't prove what would have happened, because it didn't happen. That's the whole point.

This creates a persistent organizational challenge for procurement risk management: the function's greatest successes are invisible. The CPO who invests in risk monitoring and successfully avoids a major disruption gets far less recognition than the CPO who navigates a disruption heroically after it hits, even though the first outcome is objectively better for the organization.

I don't have a perfect solution to this paradox. But I have a practical one: document the near-misses. Every time your risk monitoring capability detects a signal early and enables a proactive response, document what happened, what the system flagged, how the team responded, and what the estimated impact would have been if the disruption had not been anticipated. Build a portfolio of these near-miss case studies over time. When leadership asks "what's the ROI of our risk monitoring investment?" you have a collection of specific, documented examples of value delivered—not savings captured, but losses avoided, which in a world of increasing supply chain volatility is often worth far more.

The Japanese earthquake. The pandemic. The Suez Canal. The semiconductor crisis. The energy disruptions. Each of these events was, for most organizations, a surprise that exposed the brittleness of their supply chains and the limitations of their risk visibility.

The question is not
whether the next disruption is coming.
It is.
The question is
whether you'll see it coming or
whether it will surprise you.

AI-powered risk monitoring is, at its core, a bet that seeing around corners is possible—that the signals are there if you have the tools to detect them, and that early warning translates into better outcomes. It's a bet that the evidence overwhelmingly supports. The organizations that have made this investment are consistently more resilient, more responsive, and more prepared than those that haven't.

Build the capability. Monitor the signals. Prepare before you have to react.

Because the next disruption isn't going to call ahead.

Now let's turn from risk, the defensive capability, to offense. Let's talk about how to use AI to prepare for the conversations where value is actually created. It's time to go deep on negotiation preparation.

CHAPTER 7
THE AI-POWERED NEGOTIATION PLAYBOOK

WALKING INTO THE ROOM BETTER PREPARED THAN ANYONE EXPECTS

Quick question. Pick the most important negotiation you've got coming up in the next ninety days. The renewal that's been giving you anxiety. The new sourcing event with the supplier who's been telling everyone in your organization how indispensable they are. The contract where the stakeholders have already decided what they want and you're just there to "handle the paperwork."

Got one in mind? Good.

Now answer this. How much time are you actually going to spend preparing for it?

Be honest. I'm not asking what your organization's preparation methodology says you should spend. I'm not asking what you'd spend if your calendar weren't already on fire. I'm asking what you actually expect to spend, given the realities of your current week.

If your honest answer is somewhere between "a couple of hours the night before" and "I'll skim the supplier's last proposal during the drive over," you are not alone. You are, in fact, in the company of approximately every procurement professional I have ever known, including some of the best. The dirty secret of our profession is that the negotiations that determine whether organizations pay too much or get exactly what they need are routinely prepared for in the cracks between everything else on the calendar, because the calendar never has actual cracks, it just has fewer flaming items than usual.

This is the chapter where that changes.

Not because I'm going to give you more time. I can't do that. (If I could, I'd be running a very different business.) Because preparation that used to take a team of analysts two weeks of calendar time can now be done by one procurement professional in an afternoon. Comprehensive market intelligence. Detailed supplier financial analysis. Cost driver modeling. Competitive landscape mapping. Contract terms benchmarking. Creative deal structure generation. The full preparation dossier that great negotiators have always built when they had the resources to build it—and that the rest of us have always wished we had time for—is now buildable in a focused four-hour session by a single procurement professional with the right approach and an AI tool.

That is not a small change. That is the change.

Everything we've covered in Part Two so far converges here. The spend visibility from Chapter Four tells you what you're actually buying. The market intelligence from Chapter Five tells you what the market actually looks like.

The supplier risk monitoring from Chapter Six tells you which way the wind is blowing. All of it—every capability we've built across this Part—exists for one reason. To make you the most prepared person in the room when the conversation that determines outcomes actually happens.

Let's build the playbook.

A note for the senior leaders:

If you're trying to decide where to start your AI investment in procurement, I'll save you the suspense: start here.

Negotiation preparation is the application where AI produces the most immediately measurable, directly attributable financial return, the easiest business case to make, the fastest value to realize, and the most convincing proof point for broader adoption. The math is simple enough for any CFO. Better-prepared negotiators consistently produce better outcomes.

AI makes thorough preparation feasible across a portfolio rather than just the marquee deals. Apply that uplift across fifty negotiations a year on hundreds of millions of contract value and the cumulative impact pays for everything you'd want to spend on AI capability several times over. The strategic question is not *whether*. It is *how fast* you can build the capability before the competitors reading this same book beat you to it.

WHY NEGOTIATION PREPARATION IS THE HIGHEST-ROI APPLICATION OF AI IN PROCUREMENT

I've made the case throughout this book that AI amplifies procurement expertise rather than replacing it. Nowhere is that amplification more direct, more immediately measurable, and more clearly tied to financial outcomes than in negotiation preparation.

Here's the math, and it's simple enough that even a CFO who's skeptical about AI investments can follow it. In my experience, well-prepared negotiators who enter a negotiation with comprehensive market intelligence, deep supplier insight, clear alternatives, and creative options consistently produce outcomes meaningfully better than equivalently skilled negotiators who are less prepared. The improvement on a $5 million contract typically runs into six figures. On a $50 million engagement, it runs into seven. Across a procurement portfolio of $500 million, even assuming you're only negotiating a portion of that spend in any given year, the preparation-driven value improvement runs well into the millions.

Now consider the cost of that preparation. In the old model, thorough preparation for a major negotiation required a team of two to three people working for one to three weeks—call it forty to a hundred person-hours of research, analysis, and synthesis. At fully loaded labor costs, that's $5,000 to $15,000 per negotiation in team time alone, plus whatever you're paying for market intelligence subscriptions, analyst reports, and other research inputs.

In the AI-augmented model, one procurement professional can produce comparable, and often superior, preparation in two to four hours. The cost of the AI tools is negligible relative to the value they produce.

The constraint is no longer time or resources. It's the procurement expertise required to ask the right questions, direct the right analysis, and translate the output into negotiation strategy.

This is why I tell every procurement leader I work with that if they're going to start their AI adoption journey in one place, they should start here. Not because spend analysis isn't important—it is. Not because risk monitoring isn't valuable—it absolutely is. But because negotiation preparation is the application where AI produces the most immediately measurable, directly attributable financial return. It's the easiest business case to make, the fastest value to realize, and the most convincing proof point for broader AI adoption across the procurement function.

THE COMPLETE PREPARATION FRAMEWORK: SEVEN PHASES

Let me walk you through the complete AI-powered negotiation preparation framework that I've developed, refined, and used across dozens of negotiations over the past two years. This framework integrates the market intelligence workflow from Chapter Five, the risk assessment capabilities from Chapter Six, and the negotiation principles from Chapter Two into a single, repeatable system.

I'm going to use a running example throughout this

section: a contract renewal with a mid-market IT staffing provider. Current contract value is $3.2 million annually. The supplier has been in place for four years and they've signaled a price increase of six to eight percent at renewal. Your stakeholders are generally satisfied with the service but frustrated by some recent quality issues with the candidates being submitted. The contract expires in seventy-five days.

This is a bread-and-butter procurement negotiation—not the nine-figure strategic sourcing program, but the kind of mid-market renewal that represents the majority of a category manager's negotiation workload. It's exactly the kind of negotiation where AI preparation creates the most impact, because in the old model, a $3.2 million contract often didn't justify the level of preparation that would have produced the best outcome. It was important enough to negotiate seriously, but not important enough to warrant two weeks of dedicated research.

So the category manager would do what they could in the time available: pull last year's benchmarks, review the supplier's recent performance data, have a quick conversation with the stakeholders, and walk into the negotiation with preparation that was good but not great.

AI makes great preparation feasible for every negotiation, not just the marquee ones. And that's where the cumulative value across a portfolio becomes transformative.

Phase One: Define Your Objectives and Constraints (30 minutes, mostly human work)

Before you touch an AI tool, you need to be clear about what you're trying to accomplish. This phase is almost entirely human judgment—it's the strategic framing that directs everything that follows.

What are your must-haves versus your nice-to-haves?

In our IT staffing example, a must-have might be no price increase above three percent (your budget constraint). A nice-to-have might be improved service levels on candidate quality. Another nice-to-have might be a dedicated account manager rather than the rotating team you've been dealing with.

What is your BATNA?

What happens if you don't renew with this supplier? Do you have alternatives qualified and ready? How disruptive would a transition be? How long would it take? In our example, let's say you have one alternative partially qualified from a competitive assessment last year, and a transition would take approximately ninety days with moderate disruption. That's a credible but not overwhelming BATNA—strong enough that walking away is a real option, but not so strong that you can afford to be cavalier about the relationship.

What is your reservation point? The worst deal you'd accept before walking away?

In our example, maybe it's a five percent increase with no service level improvements—anything worse than that, and you're better off transitioning to the alternative.

What value dimensions are you willing to trade across?

Contract length? Volume commitments? Payment terms? Scope expansion? Exclusivity? Map these out before you start your research, because they'll guide the creative options you ask the AI to generate.

What are your stakeholders' priorities? This connects directly to the stakeholder alignment discussion from Chapter Two. What does the business care most about—price, quality, flexibility, innovation? Where are they willing to compromise, and where will they push back? In our example, the stakeholders have told you that candidate quality is their biggest frustration, and they'd accept a moderate price increase if it came with measurable quality improvements. That's valuable information that shapes your entire negotiation strategy.

Write all of this down. Not in your head—on paper or in a document. This becomes the strategic brief that guides your AI-powered research and ensures that the intelligence you gather is directed at answering the questions that actually matter for this specific negotiation.

Phase Two: Market Intelligence (45 minutes)

Now you deploy the market intelligence workflow from Chapter Five, calibrated specifically for this negotiation. Here's how I structure it for a staffing category renewal.

Market overview. Ask your AI tool to analyze the current IT staffing market—specifically the segment and geography relevant to your contract. What are the bill rate trends? How tight is the labor market in

the skill sets you're buying? What's the competitive landscape look like—how many credible mid-market staffing providers operate in your geography? Are there any market-level trends that would support or undermine the supplier's case for a price increase?

In our example, the AI might reveal that IT staffing bill rates in your metro area have actually been declining modestly as the post-pandemic IT hiring surge has cooled and remote work has expanded the available talent pool geographically. That's a direct contradiction of the supplier's "costs are up" narrative, and it's the kind of market intelligence that reshapes a negotiation.

Competitive alternatives.

Ask the AI to identify and profile the most credible alternative providers. For each one: their market position, geographic coverage, specialization, typical pricing model, and any available intelligence about their win rates against your current supplier. If you have the partially qualified alternative from last year, has anything changed about that company's capabilities, market position, or pricing? Ask the AI to update the profile.

Cost driver analysis.

Ask the AI to deconstruct the cost structure of IT staffing services in your market. What are the key components? Base labor costs (which track wage indices), recruiter overhead, benefits and employment costs, margin. For each component, what's the recent trend? This analysis lets you evaluate the supplier's price increase proposal on its merits—not just "that seems high" but "here's specifically where your proposed increase exceeds what the underlying cost drivers would justify."

Output: A market context document that you understand well enough to reference naturally in conversation and that provides the factual foundation for your negotiation positions.

Phase Three: Supplier Intelligence (30 minutes)

Now you build the supplier-specific dossier, following the approach from Chapter Five.

Financial analysis.

If the supplier is publicly traded, analyze their recent financial performance. If they're privately held—which many mid-market staffing firms are—use the proxy indicators we discussed: LinkedIn employee count trends, office location changes, news coverage, and any available financial data from credit services or industry databases.

What you're looking for: Signals that inform your negotiation leverage. Is this supplier growing or struggling? Are they winning or losing market share? Are they investing in new capabilities or cutting costs? Are they dependent on your account—would losing it be a meaningful blow, or are you a small piece of their portfolio?

In our example, the AI might reveal that the supplier

recently lost a large account to a competitor—information scattered across a LinkedIn post from the departing account manager and a brief mention in a staffing industry newsletter. That loss means the supplier has capacity to fill and extra motivation to retain your business—leverage you wouldn't have known about without the research.

Account team intelligence.

Who will be at the table? What's their background? How long have they been in their role? In staffing, account manager tenure is a meaningful signal—a new account manager might be more eager to make concessions to establish the relationship, while a long-tenured one might have more authority and more attachment to existing pricing structures.

Service performance analysis.

This one uses your internal data rather than external research, but AI can help synthesize it. Upload your performance tracking data—hire quality metrics, time-to-fill statistics, candidate rejection rates, stakeholder satisfaction scores—and ask the AI to identify trends. Where has performance been strong? Where has it been declining? Where are the specific, documented performance gaps that justify your push for service level improvements?

This analysis serves a dual purpose. It informs your negotiation strategy, but it also gives you documented, data-backed evidence to support your positions in the negotiation. "We've noticed some quality challenges" is a vague concern. "Your candidate acceptance rate has declined from 78% to 61% over the past three quarters, with the decline concentrated in your senior developer submissions" is a specific, evidence-based performance issue that the supplier has to respond to substantively.

Output: A supplier dossier covering financial position, account team profile, and performance trends, with the negotiation implications of each highlighted.

We're three of seven in. If this is starting to feel like a lot of preparation for one negotiation, you're right. The first time you do it this way takes a week of part-time work to build the muscle. The fifth time takes a focused afternoon. After a while you've stopped thinking about it as a process and started thinking about it as how you prepare.

Phase Four: Risk Assessment (15 minutes)

Using the risk monitoring approach from Chapter Six, conduct a focused risk assessment of both your current supplier and your alternatives.

For the current supplier: are there any financial, operational, or reputational risk signals? Is the company stable enough that a multi-year commitment makes sense? Are there concentration risks—is this supplier also providing staffing to your competitors, which could create conflicts of interest or capacity constraints?

For the alternatives: what are the transition risks? How long would onboarding take? What's the probability of service disruption during a transition? What's the alternative supplier's risk profile—are they financially stable, operationally capable, and reputationally sound?

This phase typically takes the least time because it leverages the monitoring workflow you should already have in place from Chapter Six. If you've been doing monthly risk scans on your critical suppliers, you already have a current risk picture and just need to refresh it for the negotiation.

Output: A brief risk assessment covering both the current supplier and alternatives, with the negotiation implications noted.

Phase Five: Contract Analysis (30 minutes)

This phase uses NLP-powered contract analysis to extract intelligence from your current agreement and benchmark it against market standards.

Upload your current contract to your AI tool—or if you're using a contract intelligence platform, pull the extracted terms from there. Ask the AI to identify the key commercial terms: pricing structure, escalation provisions, termination and notice requirements, auto-renewal clauses, service level definitions and remedies, liability limitations, and any other provisions that affect the economic or risk profile of the agreement.

Then ask the AI to benchmark these terms against what's typical in the market. Where are your terms more favorable than standard? Where are they less favorable? Are there provisions that are missing entirely that you should be negotiating for?

In our IT staffing example, the contract analysis might reveal that your current agreement doesn't include a most-favored-customer clause—a provision that guarantees you pricing at least as favorable as what the supplier offers to comparable clients. It might reveal that your termination-for-convenience provision requires ninety days' notice, which is longer than the sixty-day market standard. And it might reveal that the auto-renewal clause triggers in forty-five days—which means if you don't actively renegotiate, you're locked in for another year at whatever terms the supplier proposes.

That auto-renewal discovery alone could be worth tens of thousands of dollars—because if you miss the window, you've lost your negotiating leverage for an entire year. This is exactly the kind of detail that gets buried in contract language and overlooked by procurement professionals who are too busy to read their contracts carefully. The AI catches it because it reads every clause, every time, without fatigue or distraction.

Output: A contract analysis highlighting key terms, benchmarking gaps, and time-sensitive provisions that affect your negotiation timeline and strategy.

Phase Six: Creative Options Generation (20 minutes)

This is the phase that most directly connects to the principled negotiation framework from Chapter Two—the practice of expanding the pie before dividing it by identifying creative options that create value for both parties.

Armed with everything you've gathered in phases two

through five (market intelligence, supplier insight, risk assessment, and contract analysis) you now ask the AI to help generate creative deal structures. The prompt should reference the specific information you've uncovered.

Here's an example of how I'd structure this prompt for our IT staffing scenario:

"Based on what we've discussed, I'm negotiating a renewal with a mid-market IT staffing provider. Here's what I know: market bill rates in our geography are declining slightly. The supplier recently lost a major account and has capacity to fill. Their candidate quality has been declining, with acceptance rates dropping from 78% to 61%. My stakeholders value quality improvement over price reduction. I have a partially qualified alternative. The current contract lacks a most-favored-customer clause and has an auto-renewal window closing in 45 days.

Generate seven creative deal structure options that could create value for both parties. Each option should address at least two of the following: price, quality, contract length, volume commitments, scope, service levels, and risk allocation. For each option, explain what we're giving the supplier and what we're getting in return, and assess the likelihood that the supplier would find it attractive given their current situation."

The AI will generate options that a time-pressed category manager might never have considered. Perhaps a two-year extension (addressing the supplier's need for revenue stability after losing that major account) in exchange for a dedicated quality team and a price hold or modest reduction. Or a gain-share model where the supplier is incentivized financially for improving candidate acceptance rates above an agreed threshold. A volume commitment with a price tier structure that gives the supplier predictability while giving you lower rates if volumes exceed certain levels. An expanded scope, adding a new skill category that the supplier can serve, in exchange for preferential pricing across the entire engagement.

Not all of these options will be practical. Some will be naive about the operational realities of your situation. But expanding the pie by generating a wide range of creative options before narrowing to the ones that make sense consistently produces better negotiation outcomes than showing up with a single position and fighting over it.

Output: A menu of creative deal structure options, each with a clear value proposition for both parties.

Phase Seven: Strategy Synthesis and War-Gaming (30 minutes)

The final preparation phase brings everything together into a negotiation strategy and then stress-tests it through war-gaming—using the AI to role-play the supplier's likely responses.

Strategy synthesis.

Ask the AI to synthesize all the intelligence gathered in phases two through six into a negotiation preparation brief. The brief should include: your three strongest leverage points (with the evidence to support each one), the supplier's likely opening position and the arguments they'll use, your recommended opening position with market data to support it, the two or three creative options you plan to propose, the concessions you're prepared to make (and what you expect in return for each one), and the walk-away point if the negotiation reaches impasse.

Review this brief critically. Does it account for the organizational context that only you know, reflect your stakeholders' priorities accurately, address the relationship dynamics that the AI can't see? Layer in your judgment. The AI provides the analytical foundation. You provide the strategic intelligence that transforms analysis into a plan that will actually work in your specific organizational context.

War-gaming.

This is one of the most powerful and underutilized applications of generative AI in negotiation preparation. Ask the AI to role-play the supplier's side of the negotiation. Tell it: "Act as the account executive for this IT staffing provider. You've been told by your VP that you need to get at least a five percent increase on this renewal. The client is important but not your largest. You know your candidate quality has been slipping but you believe it's due to the tight labor market, not your processes. How do you respond to each of the following positions from the buyer?" Then present your planned positions and see how the AI responds.

The AI's responses won't perfectly replicate what the real supplier will say. But they'll surface the obvious counterarguments, help you stress-test your positions, and prepare you for the pushback you're most likely to encounter. I have been genuinely surprised by how often the AI's war-gaming produces arguments that the supplier's team subsequently makes in the actual negotiation—sometimes using strikingly similar language. It turns out that experienced sales teams and AI systems drawing on large datasets of sales methodology arrive at similar playbooks. Which means that preparing against the AI's counterarguments is excellent preparation for the real conversation.

Output: A complete negotiation strategy document and the confidence that comes from having stress-tested your approach against a capable adversary—even if that adversary is artificial.

Total preparation time across all seven phases: approximately three to four hours.

Compare that to the traditional preparation timeline of one to three weeks for comparable depth, and the transformation is clear. But the more important comparison isn't time—it's coverage. In the traditional model, time constraints meant that most of these phases were either skipped or executed superficially. The market intelligence was based on last year's benchmarks. The supplier analysis was gut feel. The contract review was a skim. The creative options were whatever the category manager could think of in the shower that morning. And the war-gaming didn't happen at all.

In the AI-augmented model, every phase gets the depth it deserves—even for a $3.2 million contract that wouldn't have justified this level of preparation in the old world.

ADAPTING THE FRAMEWORK: DIFFERENT NEGOTIATIONS, DIFFERENT PREPARATION

The seven-phase framework I've described is the full preparation treatment. But not every negotiation justifies the full treatment—and part of the professional judgment that distinguishes great procurement teams from good ones is the ability to calibrate preparation depth to the stakes, complexity, and strategic importance of each negotiation. There are two dimensions to that calibration: how much preparation depth to invest, and how to adapt the framework for different negotiation types.

Calibrating by Stakes: A Preparation Maturity Model

Here's the maturity model I use to determine how much preparation each negotiation warrants.

Level One: Quick Prep (30-45 minutes). For routine renewals under $500,000 with stable suppliers, limited complexity, and low strategic stakes. Quick market rate check, contract terms review for auto-renewal and escalation provisions, brief scan of supplier's current situation. The goal isn't comprehensive intelligence—it's ensuring you don't miss an obvious opportunity or walk into a deadline trap.

Level Two: Standard Prep (2-3 hours). For mid-market negotiations between $500,000 and $5 million where the relationship is ongoing, the category is understood, and the stakes are meaningful but not transformative. Phases two through five of the framework, executed at moderate depth. Market overview, supplier financial quick scan, contract analysis, and two to three creative options. Thanks to AI, this preparation level is now feasible in an afternoon rather than a week, and every negotiation above a half-million dollars should receive it.

Level Three: Full Prep (4-6 hours). For significant negotiations above $5 million, complex multi-dimensional deals, strategic supplier renewals, or situations where competitive dynamics or relationship challenges create elevated stakes. The complete seven-phase framework executed thoroughly, including war-gaming and stakeholder alignment preparation. This is the preparation that, applied across the portfolio, generates millions in incremental value.

Level Four: Deep Prep (2-3 days). For the highest-stakes negotiations—nine-figure strategic sourcing programs, company-critical supplier relationships, competitive selections that will define the organization's operating model for years. Everything in Level Three, plus extended market analysis covering multiple geographies and sub-segments, deep financial modeling of deal structure options, comprehensive legal review, executive stakeholder alignment sessions, and multiple rounds of war-gaming against different supplier team compositions and strategies.

The judgment about which level each negotiation deserves remains human. AI doesn't change the maturity model—it shifts every level to a faster timeline. What used to be a Level Three preparation requiring two weeks is now achievable in four to six hours. What used to be Level Four requiring a month is now achievable in two to three days.

The portfolio impact of this acceleration is the real story. In the pre-AI era, most procurement organizations could only execute Level Three or Level Four preparation for their top five to ten negotiations per year. The remaining forty or fifty negotiations received Level One or Level Two preparation at best—or, frequently, no structured preparation at all. The category manager walked in with their experience, their relationship knowledge, and their gut feel. Sometimes that was enough. Often it wasn't. In the AI-augmented era, a procurement team can execute Level Two preparation for every negotiation above $500,000 and Level Three preparation for every negotiation above $5 million. That expansion (applying structured, intelligence-based preparation to forty additional negotiations per year that previously received minimal preparation) is where the cumulative financial impact becomes transformative.

Calibrating by Negotiation Type

The seven-phase framework is calibrated for a contract renewal—the most common negotiation scenario in procurement. But not every negotiation is a renewal, and the preparation framework needs to flex for different contexts. Let me walk through the most common variations.

New Sourcing: The Competitive Bid

When you're running a competitive sourcing event, evaluating multiple suppliers for a new contract rather than renewing with an incumbent, the preparation emphasis shifts significantly.

Market intelligence becomes even more important, because you're not just understanding one supplier's position—you're mapping an entire competitive landscape. The AI-powered market analysis from Phase Two should be broader: covering more suppliers, more deeply, with explicit attention to how each supplier's capabilities and pricing compare to the others. Ask the AI to identify the key differentiators between your shortlisted suppliers, not just their individual strengths and weaknesses.

Supplier intelligence in Phase Three should be conducted on every shortlisted supplier, not just the incumbent. This is where AI's speed advantage is most pronounced. Building a detailed dossier on one supplier takes thirty minutes. Building comparable dossiers on five shortlisted suppliers using traditional methods would take a week or more. With AI, it takes about two and a half hours—thirty minutes per supplier, plus synthesis time. That comprehensive view of all competitors' positions, pressures, and priorities gives you an information advantage that transforms the sourcing process.

The creative options phase becomes less relevant in a competitive bid—the competition itself creates value through market pressure, and the primary negotiation vehicle is the RFP process rather than one-on-one creative deal structuring. Instead, redirect that time toward evaluation criteria development: use the AI to help you design evaluation criteria and scoring methodologies that capture the dimensions of value that matter most for your specific situation, rather than defaulting to a generic price-weighted scorecard.

The Sole-Source Renewal: When You Have No Alternative

This is the scenario we explored in Chapter Two with the life sciences SaaS vendor—the negotiation where your BATNA is essentially zero and competitive leverage doesn't exist. It's also the scenario where AI-powered preparation is most critical, because when you can't create leverage through competition, you have to create it through intelligence, creativity, and stakeholder alignment.

Phase Two (market intelligence) should focus less on competitive alternatives (since, by definition, there aren't meaningful ones) and more on understanding the supplier's market position, growth strategy, and commercial pressures. What's happening in their industry that creates opportunities for creative deal structures? Maybe they're under pressure to show customer retention metrics, launching new products that need adoption, or trying to expand into adjacent markets where a case study from your industry would be valuable. Each of these is a potential source of leverage that doesn't depend on competitive threat.

Phase Three (supplier intelligence) becomes the single most important phase. In a sole-source renewal, your leverage comes from understanding the supplier's situation better than they expect you to, and from using that understanding to propose deal structures that serve both parties' interests. The financial analysis, the account team assessment, and the strategic priorities research should be as deep as you can make them. This is where the ninety-minute dossier might become a three-hour dossier—and the extra time is justified by the stakes.

Phase Six (creative options) becomes the centerpiece of your preparation. Without competitive leverage, the negotiation is won or lost on the creativity of your deal structures. The AI is particularly valuable here because it can generate options you haven't considered—combinations of contract length, scope, volume commitment, payment terms, case study participation, and product adoption that create value for the supplier in dimensions you might not have thought of, in exchange for the commercial concessions you need.

And Phase Seven's war-gaming becomes essential. In a sole-source negotiation, the supplier knows you don't have alternatives, and they'll test your positions accordingly. War-gaming against the AI (having it play the role of a confident, well-prepared account executive who knows you're captive) prepares you for the pushback you'll face and helps you develop responses that maintain your position without relying on competitive threats you can't credibly make.

The Troubled Relationship: When Performance Has Degraded

Sometimes you're not negotiating a contract renewal—you're negotiating a course correction with a supplier whose performance has deteriorated to the point where the relationship needs to be reset. This requires a different preparation emphasis.

Phase Three should include a comprehensive, data-backed performance analysis—not generalities about "quality has been slipping," but specific, documented, time-series data showing exactly how performance has degraded, in which areas, by how much, and since when. Ask the AI to analyze your performance tracking data and produce a fact-based narrative of the performance trajectory. This analysis becomes the foundation of your negotiation—it's the evidence that supports your demand for change.

The creative options phase should include remediation structures: gain-share arrangements tied to performance improvement, step-up service level requirements with commercial consequences for failure, and transition planning frameworks that establish clear milestones and off-ramps if the remediation doesn't work. These aren't punitive measures—they're accountability structures that give the supplier a clear path to recover the relationship while protecting your organization if they can't.

The risk assessment phase should include a detailed analysis of what happens if the relationship doesn't recover and you need to transition. This contingency planning isn't about threatening the supplier—it's about knowing, with confidence, whether you can afford to walk away if the remediation fails. That knowledge affects your negotiation posture in ways that are subtle but powerful. The category manager who knows they can transition if necessary negotiates very differently from the one who is terrified of the disruption.

THE WAR-GAMING DEEP DIVE: A DETAILED EXAMPLE

I want to expand on the war-gaming concept from Phase Seven with a detailed example, because this capability is so underutilized and so powerful that it deserves more attention.

The prompt I use to kick off a war-gaming session:

"Act as the [title—e.g., VP of Sales, Account Executive, Regional Director] for [type of supplier company]. You've been given the following objectives by your leadership: [describe what you know or infer about the supplier's priorities—e.g., retain the account, achieve a minimum X% price increase, get adoption of a new product, protect margin, etc.]. You know the following about the buyer's situation: [describe what the supplier likely knows about you]. I'm going to present our negotiation positions. Respond exactly as you would in a real negotiation—defend your positions, identify weaknesses in mine, propose counter-offers, and use whatever leverage you reasonably have. Be tough but realistic."

Then start the role-play. The full prompt with follow-up sequences is in Appendix A.

The scenario: you're preparing for a renewal negotiation with a cloud infrastructure provider. Your annual spend is $4.5 million. The supplier has proposed a twelve percent increase, citing increased energy costs and infrastructure investment. You've done your preparation: market rates suggest a two to four percent increase is appropriate, the supplier recently lost their largest customer in your region, and their earnings call transcript reveals that management is under pressure to demonstrate customer retention to skeptical investors.

Here's how the war-gaming conversation with your AI tool might unfold:

You: "Act as the VP of Sales for this cloud infrastructure provider. You've been told by your CEO that customer retention is the top priority this quarter. You've just lost a major account in this region. Your CFO has set a floor of eight percent increase for all renewals to protect margins. You know the buyer has been talking to one of your competitors but you're not sure how serious they are. I'm going to present our counter-proposal—a two-year extension with a three percent increase in year one and a two percent increase in year two, with enhanced SLAs and a commitment to migrate two additional workloads to your platform. Role-play your response."

AI (as supplier VP): "I appreciate the commitment to a two-year extension—that tells me you see long-term value in our partnership, and I do too. But I want to be transparent: a three percent increase doesn't cover our cost position. We've invested heavily in security infrastructure and sustainability initiatives that directly benefit your workloads, and our energy costs in this region have increased substantially. I can work with my team on the eight percent we proposed, and I might have some room below that, but three percent is a number I simply can't take back to my leadership. The additional workload migration is interesting—can you tell me more about the scope and timeline? And I have to ask—I've heard you've been evaluating alternatives. Is that something I should be concerned about, or is this a partnership conversation?"

You: "That's a reasonable response. Now tell me: what's the weakest part of my counter-proposal from your perspective, and what would strengthen my position?"

AI: "The weakest part is the three percent number itself—it's so far below what I need that it risks feeling like a non-serious opening rather than a genuine negotiation position. If I were advising the buyer, I'd open at four percent instead of three—still well below my floor, but credible enough that I engage rather than dismiss. Your strongest card is the workload migration commitment—that's new revenue for me that I can use to justify internal flexibility on the rate. I'd lead with that rather than the price number. And if you want to really shift the dynamic, mention the competitor evaluation early and specifically—'we've completed a technical assessment with [competitor name] and the results were positive' hits much harder than a vague reference to 'exploring options.'"

This kind of exchange (where the AI plays the supplier role, exposes the weaknesses in your approach, and suggests improvements) is valuable preparation. It's not a perfect simulation of the real conversation. But it exposes blind spots, surfaces counterarguments you need to prepare for, and refines your approach in ways that make the actual negotiation significantly more effective.

I typically run two to three war-gaming rounds before a significant negotiation—each one refining my approach based on what the previous round revealed. By the time I walk into the actual meeting, I've already heard and responded to most of the arguments the supplier will make. Not because the AI predicted exactly what they'd say, but because the AI identified the logical counterarguments to my positions—and the real supplier's counterarguments, while phrased differently, follow the same logical structure.

The procurement professionals who learn to war-game their negotiations using AI are going to have an almost unfair advantage at the table. They will have practiced the conversation before it happens. They will have refined their arguments against the best counterarguments the other side can produce. And they will walk into the room with the quiet confidence that comes from having already done this—not for the first time, but for the third or fourth time, with the AI playing the role of a progressively more challenging counterparty.

IN THE ROOM: HOW AI PREPARATION CHANGES EXECUTION

I've focused this chapter on preparation because, as I said in Chapter Two, negotiation is ninety percent preparation. But preparation also changes execution (the way you behave at the table, the confidence you project, the flexibility you can exercise) in ways that are worth naming explicitly.

The Confidence Effect

When you walk into a negotiation having done the level of preparation this chapter describes, something happens to your presence at the table that is hard to quantify but impossible to miss. You are genuinely calm because you know what you know. You've already mapped the range of likely outcomes, anticipated the supplier's arguments, and identified your walk-away point—and you know (genuinely know, not just theoretically) that walking away is a viable option because you've researched the alternatives and assessed the transition risk.

That calm confidence is visible. The supplier's team reads it. And it affects their calculations about how hard to push, how aggressive to be, and how quickly to move toward concessions. The procurement professional who walks in under-prepared, relying on toughness and bluffing to compensate for a lack of information, broadcasts uncertainty through a thousand micro-signals: the moment of hesitation when challenged on a market claim, the overly aggressive posture that signals insecurity rather than strength, the inability to respond substantively when the supplier presents data that contradicts their position.

Preparation produces confidence. Confidence produces presence. Presence produces outcomes. The AI didn't change the negotiation dynamics. The preparation did. The AI just made the preparation possible in a timeframe that makes it feasible for every negotiation, not just the ones that can justify weeks of dedicated research time.

The Flexibility Advantage

Thorough preparation also gives you something that under-prepared negotiators lack: the ability to adapt in real time. When you've mapped the value dimensions, generated creative options, and war-gamed the supplier's likely responses, you can pivot during the negotiation without losing your strategic coherence. If the supplier rejects your first creative option, you have four more ready to propose. If they present an argument you hadn't considered, you can evaluate it against the comprehensive market intelligence you've gathered rather than simply reacting emotionally. If they make a concession on one dimension, you can immediately assess its value against your preparation and decide whether to reciprocate, pocket it, or push for more—because you've already mapped the relative value of each dimension.

I've experienced this flexibility advantage firsthand dozens of times, but one instance stands out. I was in a technology contract negotiation where the supplier unexpectedly offered to include a new module they were trying to get adoption on, which they proposed adding at no cost if we'd commit to a case study. Because I'd researched the module as part of my preparation (I knew its market value, I knew why they were eager for adoption, and I'd already assessed whether it was relevant to our operations) I could evaluate the offer in real time rather than asking for a recess to think about it. I knew the module was worth approximately $200,000 annually at list price. I knew the case study commitment was low-cost for us. And I knew that accepting the module strengthened my position on the pricing discussion because it gave me a concession to point to—"we're adopting your new module, we're doing a case study, and we need the pricing to reflect the partnership we're building."

The ability to evaluate an unexpected offer immediately and integrate it into my negotiation strategy without missing a beat, was the direct product of preparation. The AI had given me the background on the module as part of the supplier intelligence phase. Without that preparation, I would have had to say "interesting, let me think about that," which would have slowed the negotiation's momentum and signaled that I wasn't fully prepared.

Real-Time AI Support During Negotiations

An emerging practice worth mentioning briefly: using AI tools in real time during negotiations, not just in preparation. This might look like having a generative AI tool open on your laptop during a session, ready to quickly analyze claims the supplier makes, check market data they cite, or evaluate the financial implications of a proposal they put on the table. The supplier proposes a three percent annual escalator? You can confirm whether that's above or below comparable agreements in thirty seconds without interrupting the conversation flow.

The etiquette here matters: do this transparently. Having someone on your team openly taking notes and checking data on a laptop is normal in modern negotiations. Having them secretly using AI to fact-check the supplier's statements without their knowledge feels deceptive and is inconsistent with the ethical framework we discussed earlier. Be straightforward about it. Most sophisticated suppliers will appreciate the transparency and may even be doing the same thing themselves.

THE PORTFOLIO VIEW: SCALING AI-POWERED PREPARATION ACROSS YOUR ORGANIZATION

Everything I've described so far has been about preparing for a single negotiation. But the real transformative potential of AI-powered preparation reveals itself when you scale it across an entire procurement portfolio.

Consider a procurement organization that manages three hundred active supplier relationships, of which approximately fifty will go through some form of negotiation or renewal in a given year. In the traditional model, maybe ten of those fifty negotiations receive thorough preparation—the biggest, most strategic, most politically important ones. The other forty get lighter-touch preparation constrained by the team's bandwidth. Some get minimal preparation at all—the category manager walks in with last year's pricing, a vague sense of the market, and hopes for the best.

In the AI-augmented model, every one of those fifty negotiations can receive the full seven-phase preparation treatment in three to four hours. That's two hundred hours of preparation work for the full portfolio—a meaningful investment, but well within the capacity of a typical procurement team. And the return on that investment is substantial: even modest improvements in negotiation outcomes, applied to fifty negotiations across hundreds of millions of dollars in contract value, produce a cumulative financial impact that is transformative.

But the value goes beyond individual negotiation outcomes. When you're consistently preparing at this level across your portfolio, you're building an organizational intelligence asset. Every supplier dossier, every market analysis, every contract benchmark becomes part of a growing knowledge base that informs future negotiations. The market intelligence you gathered for this quarter's IT staffing renewal informs next quarter's renewal with a different staffing provider. The creative deal structures that worked in one category inspire similar approaches in others. The war-gaming insights, the supplier counterarguments that proved most effective, become institutional knowledge that the entire team can learn from.

Over time, this creates a compounding advantage. Each negotiation makes the next one better prepared. Each market analysis makes the next one faster and more focused. Each creative option that succeeds in one category suggests possibilities in others. The organization gets smarter about its markets, its suppliers, and its own negotiation patterns with each cycle—and AI ensures that the learning is captured, synthesized, and applied rather than locked in individual procurement pros' heads.

Pay attention to this part. Stakeholders are where deals quietly fall apart. The room with the supplier is rarely the hardest room. The hardest room is the one with your own people the day before, when you're trying to get them aligned on what "we" actually want—and discovering that we don't agree.

THE STAKEHOLDER PREPARATION: YOUR MOST IMPORTANT INTERNAL NEGOTIATION

I want to revisit the stakeholder dimension, because AI-powered preparation changes the internal negotiation as much as it changes the external one.

Remember the story from Chapter Two—the SaaS renewal where I predicted the vendor's playbook to my stakeholder, earned his trust, and then executed the coordinated good-cop-bad-cop play? The foundation of that story was credibility. My stakeholder trusted my strategy because I had demonstrated, in concrete terms, that I understood the vendor's situation better than the vendor expected us to.

AI-powered preparation makes that kind of credibility building significantly easier and more repeatable. When you walk into a stakeholder briefing with a dossier that includes the supplier's recent financial performance, their strategic priorities as articulated in their own earnings calls, the market dynamics that affect their pricing, and a set of creative negotiation options supported by data—you are projecting a level of expertise and preparation that most business stakeholders have never experienced from their procurement partners.

The effect on stakeholder engagement is profound. Stakeholders who have spent years viewing procurement as a processing function that occasionally negotiates price discounts start seeing procurement as a strategic partner with real market intelligence and analytical capability. They become more willing to invest time in pre-negotiation planning, more receptive to procurement-led negotiation strategies, and—in the language of Chapter Two—allies who amplify your leverage rather than obstacles who undermine it.

I've seen this dynamic play out repeatedly since incorporating AI into my preparation workflow. Stakeholder meetings that used to be thirty-minute check-ins where I confirmed that the contract was being renewed have become hour-long strategy sessions where the stakeholder and I are collaboratively developing a negotiation approach based on shared intelligence. The stakeholder brings organizational context and relationship history. I bring market intelligence and analytical depth. Together, we develop strategies that neither of us could have built alone.

The practical advice here is specific: when you brief stakeholders before a negotiation, share the key insights of your preparation—the market data that contradicts the supplier's narrative, the competitive alternatives that expand your options, the creative deal structures that address both the organization's interests and the supplier's needs. Share enough that the stakeholder understands the strategy, trusts the preparation, and feels like a co-creator of the approach rather than a passive observer of the procurement process.

That shared preparation creates alignment.
Alignment creates leverage.
Leverage creates outcomes.

And AI-powered preparation makes the cycle possible

at a speed and depth that transforms the procurement-stakeholder relationship from transactional to strategic.

COMMON PREPARATION MISTAKES (EVEN WITH AI)

Let me close the practical portion of this chapter with the most common mistakes I see sourcing professionals make when using AI for negotiation preparation that can undermine the advantage that thorough preparation should provide.

Preparing but not internalizing.

Having a comprehensive dossier is useless if you haven't actually absorbed the information. I've seen the best buyers produce brilliant AI-powered preparation documents and then fail to reference any of the intelligence in the actual negotiation because they hadn't taken the time to internalize the key insights. The dossier is not a reference document to consult during the meeting. It's preparation material that should be so thoroughly understood that its insights flow naturally into your conversation. Read it. Understand it. Know it well enough to discuss everything inside without looking at your notes.

Over-relying on the dossier and under-relying on judgment.

The dossier tells you what the data says. It doesn't tell you what the data means in the context of your organization, your relationship, and your stakeholders' priorities. A category manager who follows the AI's recommended strategy without filtering it through their own professional judgment is not better prepared—they're just better informed, which is not the same thing. The judgment layer is where your value lives. Don't skip it.

Failing to verify critical facts.

We've covered this in previous chapters, but it bears repeating in the negotiation context specifically: never cite an AI-generated fact in a negotiation without verifying it first. The hallucination risk, the AI generating plausible but fabricated data, is real, and the consequences in a negotiation context are severe. A single incorrect market benchmark, cited confidently and then debunked by a well-prepared supplier, can destroy hours of careful preparation work by undermining your credibility on everything else you present.

Preparing for the negotiation you want rather than the negotiation you'll get.

AI tools will dutifully help you prepare for whatever scenario you describe. If you describe a scenario where you have all the leverage and the supplier is desperate, the AI will produce a preparation plan for that scenario—even if the reality is that you're the more dependent party. Be honest with the AI about your situation. Give it the uncomfortable facts. Tell it about your weak BATNA, your stakeholder who loves the incumbent, your contract that auto-renews in thirty days. The preparation is only as good as the reality check that informs it.

Preparing too much.

Yes, this is a thing. I've seen procurement professionals spend twelve hours on preparation for a $200,000 contract renewal—preparation that was thorough and brilliant and completely disproportionate to the stakes involved. The seven-phase framework is calibrated for significant negotiations; the maturity model handles the calibration question. For smaller, more routine negotiations, a streamlined version is more appropriate. Matching the preparation depth to the stakes is the discipline that distinguishes procurement teams who use AI well from those who use it indiscriminately.

THE NEGOTIATION PREPARATION MINDSET: EARNING THE RIGHT TO BE DANGEROUS

I want to close this chapter with a thought that connects all the way back to Chapter Two's observation about curiosity —because it applies here with particular force.

The procurement professionals who are getting the most value from AI-powered negotiation preparation are not the ones who use AI as a shortcut. They're the ones who use it as an amplifier for curiosity they already have— for the supplier's business, the market dynamics, the cost drivers and the competitive landscape and the regulatory environment and the technology trends that affect their categories. They wanted to understand all of this before AI existed. They just didn't have the tools to do it at the speed and depth that was needed.

AI satisfies that curiosity at scale. It gives the deeply curious procurement professional the ability to explore markets, analyze suppliers, and generate creative options at a pace that matches the speed of their thinking rather than being constrained by the speed of their research. And the result is a level of preparation that doesn't just improve negotiation outcomes—it transforms the procurement pro's experience of the negotiation itself.

There is a version of procurement negotiation that feels like a high-stakes bluffing game—walking in with incomplete information, hoping the supplier doesn't know more than you do, relying on toughness and tactics to compensate for gaps in understanding. That version is stressful, adversarial, and produces outcomes that leave value on the table.

There is another version, the version that AI-powered preparation makes possible, where you walk in knowing more about the market, the supplier, and the range of possible outcomes than the supplier expects you to know. Where your positions are supported by evidence rather than bluster. Where your creative options reflect understanding of both parties' interests. Where your confidence comes not from a persona you're performing but from the deep preparation you've done. That version is intellectually satisfying, professionally rewarding, and consistently produces superior outcomes.

The preparation earns you the right to be dangerous. Not dangerous in the adversarial, Big Ron sense—

dangerous in the sense that you are so well-informed, so well-prepared, and so creative in your approach that the supplier's sales playbook doesn't work on you. Their practiced arguments bounce off your market data; their emotional tactics fail to rattle your evidence-based confidence; their standard pricing hits your should-cost analysis and has to defend itself on the merits rather than riding on the assumption that you won't challenge it.

That's what great negotiation preparation looks like in the age of AI. Not tougher. Not louder. Not more aggressive. Better informed, better prepared, and genuinely dangerous.

CHAPTER 8
CONTRACT INTELLIGENCE

TURNING YOUR FILING CABINET INTO A STRATEGIC WEAPON

Let me tell you about the most expensive filing cabinet in the history of corporate America.

It wasn't actually a filing cabinet. It was a shared network drive, one of those digital wastelands with folder names like "Contracts_Final_v2_REVISED_USE THIS ONE" and "Old Contracts DO NOT DELETE (Dave)", containing approximately four thousand supplier contracts for a Fortune 500 company where I was brought in to lead a procurement transformation. The contracts represented roughly $2.3 billion in annual supplier spend. They governed relationships with everyone from the company's largest technology partners to the guy who delivered bottled water to the break room.

My first week on the job, I asked a simple question:

"How many of our contracts have auto-renewal clauses that trigger in the next ninety days?"

Nobody could answer it.

Not because they were incompetent. Because answering that question would have required someone to open every single contract on that network drive, read through the terms and conditions, find the renewal provision, calculate the notification deadline based on the contract's effective date, and compile the results. For four thousand contracts. Written in different formats, by different law firms, using different terminology, some of them scanned PDFs where the text wasn't even searchable, some of them in languages other than English, and a meaningful number of them stored in locations that didn't match the folder naming convention—because someone had saved them to the wrong directory three years ago and nobody had noticed.

The question I asked should have been answerable in thirty seconds. In reality, it was essentially unanswerable without a project-level effort that nobody had the bandwidth to undertake.

Three months later, we discovered, the hard way, that eleven contracts totaling approximately $18 million in annual spend had auto-renewed during my first sixty days because nobody had flagged the notification deadlines. Eleven contracts. Auto-renewed at existing terms. Terms that, in several cases, included pricing that was well above current market rates, service levels that hadn't been updated in years, and provisions that no longer reflected the organization's current requirements. The cost of those missed renewals, the difference between what we were locked into and what we could have negotiated with adequate preparation, was conservatively estimated at $2.7 million over the renewal period.

$2.7 million, lost not because someone made a bad decision, but because nobody could find the information needed to make any decision at all.

This is the state of contract management at most organizations. And if you think I'm exaggerating for dramatic effect, I invite you to walk into any procurement office in the country, ask the CPO how many of their contracts contain most-favored-customer clauses, and time how long it takes them to answer. If they can answer at all, which most can't, it will take days or weeks, not seconds. And that single data point, which contracts guarantee you pricing parity with other customers, represents potentially millions of dollars in value that the organization may or may not be capturing, and almost certainly doesn't know.

Contracts are the legal and commercial backbone of every supplier relationship. They define what you're buying, what you're paying, what performance you're entitled to, what happens when things go wrong, and how the relationship evolves over time. They are, collectively, among the most valuable documents your organization possesses. And at most organizations, they are the least accessible, least analyzed, and least leveraged strategic assets in the entire enterprise.

AI changes the economics of contract management from "too expensive to do well" to "too expensive not to." If you lead a procurement function, the question is whether yours will capture that value or continue leaving it on the table.

Let me show you how.

WHY CONTRACT MANAGEMENT HAS BEEN PROCUREMENT'S BLIND SPOT

Before we get into what AI changes, let's understand why contract management has historically been so poorly executed—even at organizations that excel at other procurement disciplines.

The core problem is structural. Contracts are complex legal documents written in specialized language. They are created during the sourcing and negotiation process—a period of intense focus and attention—and then, in most organizations, they are filed away and rarely revisited until something goes wrong or the renewal date approaches. The negotiators who crafted the terms move on to the next deal. The procurement professionals who manage the day-to-day supplier relationship may never have read the full contract. The commercial terms that were carefully negotiated become organizational folklore—"I think our payment terms are net 45"—rather than actively managed commitments.

This creates a perverse dynamic. The organization invests significant time, expertise, and energy in negotiating favorable contract terms—the work we discussed extensively in Chapters Three and Eight—and then systematically fails to capture the full value of those terms because nobody is monitoring whether the supplier is performing to the contract's standards, nobody is tracking whether the pricing escalation provisions are being applied correctly, and nobody notices when the contract's auto-renewal clause triggers because the notification deadline passed without anyone flagging it.

I've seen this dynamic produce substantial value leakage. Suppliers billing at rates above the contracted maximum because nobody checks the invoices against the rate card. Service levels degrading below contractual thresholds without triggering the remedies that the contract provides because nobody is measuring against the contractual definitions. Price escalation clauses applying annual increases based on a CPI index that has moved in the buyer's favor, meaning the prices should have gone down, but nobody noticed because nobody was tracking the index against the contractual formula.

Each of these individual leakages might be modest. A few thousand dollars here, a few percent there. But aggregated across hundreds or thousands of contracts, the cumulative value leakage is substantial. World Commerce & Contracting—the global contracting standards body, formerly known as IACCM—has been tracking this for years. Their research shows organizations lose an average of around nine percent of annual revenue through poor contract management. Best performers hold leakage to three percent. The worst lose fifteen percent or more, particularly in complex supplier ecosystems where post-signature management is weakest. The losses come from missed obligations, uncaptured entitlements, and terms that are negotiated but never enforced. On a billion-dollar procurement spend, that's $50 million to $150 million in value that was negotiated, documented in a contract, and then never realized.

That's the filing cabinet problem. Not that the contracts don't exist. Not that the terms aren't favorable. But that the information locked inside those contracts is inaccessible at the speed and scale required to actually manage it. And until AI came along, there was no cost-effective way to solve it.

Imagine Big Ron with this kind of contract intelligence.

He read the contracts that mattered. His top ten or fifteen—the ones where the spend was big enough or the relationship strategic enough that the negotiation had been a real one. He'd memorized the price-protection clauses, the liability caps, the most-favored-customer provisions. When the supplier on one of those contracts pushed back on something he thought he was owed, he'd quote the section number from memory and watch the account rep's posture change.

The other three hundred contracts in his portfolio? Filed and forgotten. Big Ron didn't read them. Nobody did. They sat on the shared drive in folders called "Contracts_-Final_v2_REVISED_USE THIS ONE." When auto-renewals triggered, he found out when the invoice came. When suppliers under-delivered on service levels, he found out when his stakeholders complained. As for escalation clauses—when one applied a CPI increase the formula didn't actually entitle the supplier to, nobody noticed, including Big Ron, because checking would have required reading the contract, finding the formula, pulling the right index, and doing the math. For three hundred contracts. He had a job to do.

Give Big Ron contract intelligence that surfaces every

auto-renewal coming in the next six months, the most-favored-customer clauses sitting in his portfolio unenforced, the service-level provisions the supplier has been quietly missing, the escalation clauses whose math should have gone the buyer's way—and the tail of the portfolio that used to leak value would have closed up. Whether he used that visibility to ride suppliers harder on their commitments or to invest in the partnership conversations he never had time for was, as always, going to be Big Ron's call. The point is that he'd finally have a choice. Same as you.

There's a scene in The Big Short—the one where Michael Burry sits alone in his office, lights off, doing what nobody else on Wall Street had bothered to do. He reads the actual mortgage documents. Not the rating agency summaries. Not the bond-issuance materials. The thousands of pages of underlying loans. And what he finds, sitting there in the documents that anyone with access could have read, is the entire shape of the financial crisis to come. The information was public. It had always been public. The constraint wasn't access—it was that reading it would have taken weeks and nobody else thought it was worth the time. Contract intelligence is the same setup. Your organization's contracts are sitting on the shared drive right now. The obligations the suppliers owe you, the entitlements you've never collected, the escalation clauses being applied wrong, the auto-renewals about to trigger—all of it is in the documents. The constraint has been the same as Burry's was: reading every contract would have taken weeks, and nobody had the weeks. AI has the weeks. AI has the weeks compressed into hours. The question is whether your organization is going to be the one that reads what's in the documents, or the one that's still relying on the summary memory of the person who negotiated them three years ago.

THE AI TRANSFORMATION: FROM DOCUMENT STORAGE TO DECISION INTELLIGENCE

AI-powered contract intelligence transforms contract management from a document storage problem into a decision intelligence capability. Let me walk you through the capabilities and what they mean for procurement practice.

Large-Scale Contract Data Extraction

The foundational capability of contract intelligence is the ability to read contracts at scale and extract structured data from them. NLP-powered systems can ingest thousands of contracts in dozens of formats (Word documents, PDFs, scanned images, even handwritten amendments) and extract key provisions: pricing terms, payment terms, liability caps, termination provisions, renewal triggers, service levels, indemnification clauses, and dozens of other commercial and legal terms.

What used to require an army of paralegals working for months can now be accomplished in hours by an AI system that doesn't need breaks, doesn't lose focus, and applies consistent extraction logic across every document. The output is a structured database where every key provision in every contract is searchable, comparable, and analyzable.

The downstream value of this capability is hard to overstate, because once your contracts are extracted into structured data, every subsequent analysis becomes possible.

"Show me every contract with an auto-renewal clause that triggers in the next sixty days." Twenty seconds.

"Show me every contract that lacks a most-favored-customer provision." Ten seconds.

"Show me every contract with a liability cap below $1 million in our top fifty suppliers by spend." Fifteen seconds.

Questions that were essentially unanswerable in the pre-AI world become trivial.

The Year Three Problem, Solved

In Chapter Two, we discussed what I call the year three problem—the dynamic where supplier contracts often produce favorable initial terms followed by gradual value erosion as auto-renewals trigger, escalation clauses compound, and the supplier becomes embedded in the organization's operations to the point where switching costs make competitive pressure ineffective.

AI-powered contract intelligence is the structural solution to the year three problem. When every contract's renewal date, escalation provision, and auto-renewal trigger is extracted into a structured database, your renewal pipeline becomes visible at any time horizon—not just the next thirty days, but the next twelve to eighteen months. You can see every renewal coming six months out. You can plan negotiation preparation for high-stakes renewals well in advance. You can identify which auto-renewal clauses are about to trigger and intervene before the deadline passes. The information that used to be locked inside thousands of unread contracts becomes a working management tool.

This sounds almost embarrassingly basic. It is basic. And yet the number of organizations that cannot reliably answer the question "what contracts are renewing in the next quarter?" is high enough to have created the $50M-$150M value leakage problem we just discussed.

The $2.7 million in missed renewal value I described at the opening of this chapter was entirely a failure of visibility—the terms were negotiable, the market was favorable, and the procurement team had the skill to negotiate better outcomes. They just didn't know the deadlines were approaching until it was too late.

But the strategic value of the renewal pipeline goes beyond avoiding missed deadlines. Once you have visibility into your full contract renewal pipeline, twelve to eighteen months out, you can make proactive decisions about where to invest your team's time, which renewals to prioritize for competitive rebid versus direct negotiation, and how to sequence multiple negotiations in the same category or with the same supplier for maximum leverage.

Here's a concrete example. A procurement organization I worked with discovered through their contract intelligence platform that they had three separate contracts with the same IT services provider, managed by three different category managers, all renewing within a six-month window. In the old world, where each category manager managed their own renewals independently, they would have negotiated three separate renewals at three different times, each with limited leverage.

With pipeline visibility, they coordinated, combining the three renewals into a single enterprise negotiation—$12 million in combined spend rather than three separate $4 million conversations. The combined leverage produced an outcome that was materially better than any of the three separate negotiations could have achieved: an enterprise discount of eight percent, standardized terms across all three agreements, a dedicated enterprise account team, and a single governance framework instead of three separate relationship management structures. The savings from the combined negotiation versus the projected outcomes of three separate negotiations: approximately $1.1 million over the contract term.

All because the renewal pipeline visibility revealed a coordination opportunity that was invisible when each category manager was looking only at their own renewals. The individual category manager couldn't see the pattern because their view was limited to their own category. The procurement leader who had a portfolio-wide view, enabled by AI-powered contract intelligence, could orchestrate a cross-category strategy that created value the fragmented approach never would have.

The renewal pipeline also enables proactive resource allocation. If you can see that the next quarter has fifteen renewals totaling $80 million, and the quarter after that has only five renewals totaling $12 million, you can plan your team's capacity accordingly—front-loading preparation work, scheduling stakeholder briefings, and ensuring that the high-concentration quarter gets the attention it deserves rather than overwhelming the team with competing deadlines. This proactive planning capability sounds basic, but most procurement organizations can't do it because they don't have reliable visibility into their renewal pipeline beyond the next thirty to sixty days. AI-powered contract intelligence eliminates the information gap that makes reactive management the default.

Terms benchmarking across the portfolio.

AI extraction allows you to compare contract terms across your entire supplier portfolio—something that was essentially impossible when terms were locked inside individual contract documents. This benchmarking reveals patterns that inform both current management and future negotiation.

For example, you might discover that your liability caps with technology suppliers range from $500,000 to $50 million, a hundred-fold variation, with no apparent correlation to the size or risk profile of the engagement. That variation is almost certainly not intentional. It's the accumulated result of different negotiations, different legal reviewers, and different suppliers' standard terms being accepted without systematic comparison to a benchmark. Once visible, it's addressable: you can establish a standard liability framework calibrated to engagement value and risk, and you can target the outlier contracts for renegotiation at the next renewal.

Or you might discover that your payment terms with similar suppliers vary widely—some at net 30, some at net 45, some at net 60, and a few at net 90—again with no apparent strategic rationale for the variation. Since payment terms have real cash flow value, normalizing them across your portfolio can create significant financial benefit. AI makes the variation visible. Your negotiation skills—amplified by the preparation framework from Chapter Seven—capture the value.

Escalation and adjustment monitoring.

Many contracts include provisions for periodic price adjustments—escalation clauses tied to inflation indices, market benchmarks, or supplier cost changes. In theory, these adjustments should work in both directions: prices should go up when costs increase and down when costs decrease. In practice, suppliers are meticulous about applying upward adjustments and reliably forgetful about applying downward ones. And most procurement organizations don't have the bandwidth to track every escalation clause against every relevant index across hundreds or thousands of contracts.

AI does. Contract intelligence platforms can track the indices referenced in your escalation clauses, calculate what the adjustment should be based on the contractual formula, compare that to what the supplier has actually charged, and flag any discrepancies. This capability alone, catching escalation clauses that have been misapplied or applied in only one direction, routinely recovers value that has been leaking for years without anyone noticing.

At a recent live deal tear-down I ran with a client, AI-powered escalation monitoring revealed that a logistics supplier had been applying annual CPI-based increases for four consecutive years—but the CPI in the relevant geography had actually declined in one of those years, meaning the supplier owed a price reduction that they had never applied. The cumulative overcharge was approximately $340,000. Not a fortune, but not trivial either—and it was money the organization was contractually entitled to but had never claimed because nobody was tracking the index against the formula.

Multiply that kind of recovery across a portfolio of hundreds of contracts with various escalation provisions, and the value becomes significant.

Risk Pattern Detection

Beyond individual contract terms, AI-powered contract intelligence can identify risk patterns across your entire contract portfolio that would be invisible when contracts are reviewed individually but become clear when analyzed collectively.

For example:

- How many of your contracts contain unlimited liability provisions where you bear the risk?
- How concentrated is that exposure—are there a small number of suppliers where your liability exposure is disproportionately high?
- What's your aggregate exposure to force majeure claims if a major disruption hits?
- How many of your contracts lack adequate data protection provisions in an era of increasing data privacy regulation?

- How many contain change-of-control clauses that would be triggered if a supplier is acquired—and given the pace of M&A activity in some of your supplier categories, how likely is that scenario?

These portfolio-level risk questions are important for organizational risk management, but they're essentially unanswerable without the ability to analyze the full contract corpus simultaneously.

A CISO who asks "what data protection obligations have we committed to across our supplier contracts?" deserves an answer.

A CFO who asks "what's our maximum aggregate liability exposure across our supplier portfolio?" deserves an answer.

Without contract intelligence, those questions go unanswered—or worse, they get answered with guesses that underestimate the actual exposure.

AI-powered contract analysis makes these portfolio-level questions answerable for the first time. And the answers, in my experience, consistently surprise the executives who ask them—usually in ways that prompt immediate action to address the risks that the analysis reveals.

Obligation Management

Every contract contains obligations—things that both parties have committed to do. Payment obligations. Delivery obligations. Reporting obligations. Insurance maintenance obligations. Audit rights. Compliance certifications. Performance reporting requirements.

In most organizations, obligation tracking is informal at best. The procurement professional who negotiated the contract remembers the key obligations. Their successor, who inherited the supplier relationship eighteen months later, may not. The supplier's obligations are tracked to the extent that failures become visible through operational impact. The buyer's obligations are tracked to the extent that the supplier reminds you about them.

AI-powered obligation extraction and tracking changes this from an informal, memory-dependent process to a systematic, automated one.

(I realize "obligation extraction" is the kind of phrase that makes the average reader's eyes glaze over before the period at the end of the sentence. Stay with me. The capability is more interesting than the name.)

The platform extracts the obligations from each contract, assigns ownership, sets due dates, and generates reminders. Your obligation to provide a quarterly forecast to the supplier by the fifteenth of each quarter? Tracked and reminded. The supplier's obligation to provide an annual SOC 2 audit report? Tracked and escalated if not received. Your obligation to conduct an annual business review with the supplier's executive team? Tracked and scheduled.

The practical impact is less about any single obligation—though missed obligations can have real consequences—and more about the cumulative effect of systematic obligation management on relationship quality and contract compliance. When both parties consistently meet their contractual obligations, trust builds. When obligations are missed—especially when they're missed because nobody was tracking them—trust erodes. And as we established in Chapter One, trust is the foundation of supplier relationships that create long-term value.

Some of you are going to be tempted to skip this section because contract work feels like the legal team's problem, not procurement's. Read it anyway. The value leakage we just discussed is being charged to your spend whether anyone is managing it or not. If procurement isn't capturing it, sooner or later someone in finance is going to ask why.

CONTRACT ANALYSIS WITHOUT A PLATFORM: THE GENERAL-PURPOSE AI APPROACH

Not every organization has invested in a purpose-built contract intelligence platform—and for many, particularly mid-sized companies or procurement teams early in their AI adoption journey, the investment may not yet be justified. But that doesn't mean you can't leverage AI for contract intelligence. General-purpose AI tools like ChatGPT, Claude, CoPilot, and Gemini are highly capable at contract analysis when directed by a sourcing professional who knows what to ask.

Let me walk you through the practical workflow for using a general-purpose AI tool to analyze contracts.

Single Contract Deep Dive

When you need to thoroughly analyze a single contract before a renewal, negotiation, or in response to a supplier issue, upload the contract to your AI tool and work through a structured set of prompts.

Commercial terms extraction:

"Read this contract and extract all commercial terms including: pricing structure and rates, payment terms, volume commitments, minimum purchase obligations, price escalation or adjustment provisions, most-favored-customer provisions, and any financial incentives or penalties. Present them in a structured table format."

Risk provision analysis:

"Identify all risk-related provisions in this contract including: liability limitations and caps, indemnification obligations (for both parties), insurance requirements, force majeure definitions and consequences, data protection and breach notification requirements, and warranty provisions. For each one, assess whether the provision favors the buyer, the supplier, or is balanced, and flag any provisions that you would consider unusual or particularly aggressive."

Operational terms review:

"Extract all operational obligations and service level provisions including: service level definitions and metrics,

performance measurement methodology, remediation and cure periods, escalation procedures, reporting requirements, and change management provisions. Identify any provisions where the language is vague or where performance standards lack measurable thresholds."

Renewal and termination analysis:

"Identify all provisions related to contract duration, renewal, and termination including: initial term and any extension options, auto-renewal clauses and their trigger dates, termination for convenience provisions and notice periods, termination for cause provisions and cure periods, transition assistance obligations, and any provisions that affect the buyer's ability to exit the agreement. Calculate the specific dates when notice must be given to prevent auto-renewal based on the contract's effective date."

Benchmark assessment:

"Based on your knowledge of standard commercial contracting practices in [category], identify the five provisions in this contract that are most unfavorable to the buyer compared to market standard. For each one, explain what the market standard provision looks like and recommend specific language improvements for the next negotiation."

This structured prompt sequence takes about forty-five minutes to execute including the review of the output, and it produces a contract analysis that is significantly more thorough than what most procurement pros would produce by reading the contract themselves, because the AI catches provisions that a time-pressured human reviewer might skim past, and because the structured prompting ensures that no major category of provisions is overlooked.

Portfolio Comparison

You can also use general-purpose AI tools for limited portfolio analysis—not at the scale of a dedicated platform, but enough to generate useful comparative insights.

Upload two or three contracts from the same category (say, your three IT staffing supplier agreements) and ask the AI to compare them on specific dimensions.

"Compare the liability provisions across these three agreements and identify which one is most favorable to us."

"Compare the termination provisions and identify which agreement gives us the most flexibility to exit."

"Compare the pricing structures and identify which one provides the best protection against cost escalation."

These comparisons are valuable for renewal preparation. If you discover that Supplier A's contract includes a most-favored-customer clause that Supplier B's lacks, you know exactly what to negotiate for in Supplier B's renewal. If you discover that your liability cap with Supplier C is a fraction of what you've negotiated with Suppliers A and B, you know where to focus your legal team's attention.

The general-purpose AI approach won't give you the continuous monitoring, automated alerting, or enterprise-scale analysis that a dedicated platform provides. But it gives any procurement professional the ability to conduct deeper, more thorough, more systematic contract analysis than was possible without AI quickly, and at essentially no cost beyond the AI subscription they're probably already paying for.

THE CLAUSE THAT WAS WORTH $4 MILLION: A CONTRACT INTELLIGENCE WAR STORY

Let me tell you a story that illustrates why contract intelligence matters in concrete, dollars-and-cents terms that would make any CFO pay attention.

The setting: a technology services agreement with a mid-sized provider.

The contract had been in place for about five years, renewed once, and was coming up for its second renewal. The annual spend was approximately $6 million. The relationship was generally positive—good service delivery, reasonable pricing, no major issues. A classic "if it ain't broke, don't fix it" renewal that, in the old world, would have been handled with a quick pricing discussion and a signature.

As part of my contract intelligence practice, I had the AI analyze the full agreement, original contract plus the first renewal amendment, as preparation for the second renewal. The analysis flagged something that I would almost certainly have missed in a manual review: a benchmarking clause buried in the contract's service level appendix.

The clause stated that the buyer had the right, once per contract term, to commission an independent benchmarking study comparing the supplier's pricing to market rates, and that if the benchmark showed the supplier's pricing was more than ten percent above the market median, the supplier was obligated to reduce their pricing to within five percent of the median within sixty days.

The clause had been in the contract since the original agreement—negotiated by my predecessor, who had apparently been a better contract negotiator than they got credit for. But it had never been exercised. Not once in five years. Because nobody knew it was there. It was buried in an appendix that nobody had read since the contract was signed.

I exercised the clause. We commissioned the benchmarking study. The study showed that the supplier's pricing was approximately fourteen percent above the market median—well above the ten percent trigger in the benchmarking provision. Under the terms of the clause, the supplier was obligated to reduce their pricing to within five percent of the median.

The supplier was not thrilled. They argued that the benchmarking methodology was flawed. They argued that the market median didn't reflect the quality of their service delivery. They argued that the clause was intended as a protection against egregious overpricing, not as a mechanism for routine rate adjustment. All fair arguments, honestly—and we had substantive discussions about the methodology and the appropriate comparison set.

In the end, we agreed on a price reduction of approximately eleven percent—split between an immediate adjustment and a phased reduction over the following twelve months. On $6 million in annual spend, that eleven percent represented approximately $660,000 in annual savings. Over the remaining four-year contract term, the cumulative value was approximately $2.6 million—and that's before accounting for the compounding effect of the reduced base price on future escalation calculations, which added approximately another $1.4 million over the contract's projected life.

$4 million in value. From a single clause. That nobody knew existed. In a contract that had been in place for five years.

That's the filing cabinet problem in its most expensive incarnation. The clause was there. It had been negotiated, probably at some cost in other concessions, by a procurement professional who understood its value. But because the contract was sitting unread in a digital filing cabinet, the clause was never used, and the value it was designed to capture was leaking away, year after year, in pricing that was fourteen percent above market without anyone knowing.

Contract intelligence found it.

Procurement expertise activated it.

And $4 million in value that had been sitting dormant for five years was finally realized.

How many clauses like that are sitting in your contract portfolio right now, waiting to be discovered?

That's not a rhetorical question. It's a business case for contract intelligence.

THE VENDOR CONSOLIDATION TRAP: A PROBLEM ONLY CONTRACT INTELLIGENCE CAN SEE

I want to spend some time on a specific contract risk that has become one of the most significant and least discussed challenges in procurement over the past decade: the vendor consolidation trap. Because this is a problem where AI-powered contract intelligence isn't just helpful—it's the only realistic defense.

Here's the pattern. You contract with Supplier A for one service and Supplier B for a different service. Both contracts are well-negotiated, competitively priced, and appropriately structured for their respective categories. Then Supplier A acquires Supplier B. Suddenly, you have two contracts with the same company—and that company has both the leverage and the incentive to push you toward a consolidated platform, usually at pricing that is higher than what you were paying for the individual services, and on terms that increase your switching costs and reduce your future negotiating leverage.

This has happened repeatedly in the technology sector over the past decade. Major platform vendors have pursued aggressive acquisition strategies, buying complementary companies and then pressuring customers to migrate to integrated platforms—often with pricing that captures the "integration premium" while eliminating the competitive alternatives that kept pricing honest in the individual categories.

The procurement professional managing the Supplier A relationship may not even know that Supplier B was acquired until the consolidated pricing proposal lands on their desk. And by then, the leverage dynamic has already shifted—because the competitor you were using as your alternative in Category B is now owned by the same company that supplies Category A.

AI-powered contract intelligence addresses this risk by providing portfolio-level visibility that reveals the concentration patterns before they become traps. When you can see all of your supplier relationships and contract terms in a single view (including parent company relationships, subsidiary structures, and recent M&A activity) you can identify the consolidation risks proactively.

The practical response might involve several moves. Negotiating change-of-control provisions into your contracts that give you the right to renegotiate or terminate if the supplier is acquired. Maintaining competitive intelligence on alternative providers in categories where consolidation is likely. Building contractual protections against forced platform migration—provisions that guarantee your right to continue using the specific product you contracted for, at the contracted price, for the remaining term. And monitoring M&A activity in your key supplier categories so that you're not blindsided when an acquisition is announced.

All of these defensive moves require the portfolio-level visibility that only contract intelligence can provide. The category manager who sees only their own contracts can't spot the cross-category concentration building. The procurement leader who sees the full portfolio, with AI-powered analysis highlighting the emerging patterns, can intervene before the trap closes.

I've watched organizations lose millions of dollars to this dynamic because they couldn't see the consolidation pattern forming across their contract portfolio until it was too late to respond effectively. Contract intelligence makes the pattern visible. Procurement expertise determines the response.

PUTTING CONTRACT INTELLIGENCE INTO PRACTICE: THE QUARTERLY CONTRACT REVIEW

Let me make this practical with a workflow that any procurement professional can implement—even without a dedicated contract intelligence platform.

The Quarterly Portfolio Review

Every quarter, conduct a systematic review of your contract portfolio using AI tools. Here's the workflow:

Step one: renewal pipeline.

Generate a list of all contracts expiring or auto-renewing in the next one hundred eighty days. For each one, note the key dates: expiration, auto-renewal trigger, required notification deadline. Flag any where the notification deadline falls within the next sixty days—these are your urgent items that need immediate attention.

If you have a contract management system, this should be a straightforward report. If you're one of the many organizations who still manage contracts in spreadsheets, shared drives, or the memories of individual sourcing professionals, this is where AI can help. Upload your contract index (or build one by having the AI read your contracts) and ask it to extract and organize the renewal timeline.

Step two: terms audit on upcoming renewals.

For each contract in the renewal pipeline, use AI to extract and review the key commercial terms. How does the pricing compare to current market rates? Have the escalation clauses been applied correctly? Are there provisions that should be renegotiated based on how the relationship has evolved? Are there missing protections (auto-renewal opt-out provisions, most-favored-customer clauses, termination for convenience rights) that should be added in the renewal?

This review produces the specific negotiation objectives for each upcoming renewal—the list of terms to improve, provisions to add, and pricing to challenge. Combined with the negotiation preparation framework from Chapter Seven, it ensures that every renewal is an opportunity to optimize rather than a deadline to meet.

Step three: compliance check.

Review a sample of your active contracts for compliance—both the supplier's compliance with their obligations and your organization's compliance with yours. Are suppliers meeting their service level commitments? Are pricing terms being applied correctly? Are insurance and certification requirements current? Are reporting obligations being fulfilled?

AI can assist with this by comparing invoice data against contracted rates, identifying discrepancies between service level reports and contractual definitions, and flagging obligations that appear to be unfulfilled. The review doesn't need to cover every contract every quarter—a rotating sample of your most significant contracts is sufficient to identify systemic issues and catch individual problems before they compound.

Step four: risk assessment.

Review the contract portfolio for emerging risks. Have any of your suppliers been involved in M&A activity that changes the relationship dynamic? Have regulatory changes created new compliance requirements that your existing contracts don't address? Have any suppliers' financial situations deteriorated in ways that increase performance risk? Are there concentration risks, too much spend with a single supplier or in a single geography, that your contract portfolio isn't adequately protected against?

Step five: action planning.

Based on the review, develop a prioritized action list for the coming quarter.

- Which renewals need full negotiation preparation?
- Which compliance issues need to be addressed?

- Which risk items need mitigation plans?
- Which terms improvements should be targeted?

The entire quarterly review, for a portfolio of several hundred contracts, takes one to two days with AI assistance. Without AI, it would take weeks and would likely never happen because the procurement team doesn't have weeks to dedicate to a contract review exercise.

AI makes the review feasible.

The procurement professional makes the review valuable by applying the judgment, context, and strategic thinking that transforms data into action.

USING AI FOR REAL-TIME CONTRACT QUESTIONS

Beyond the structured quarterly review, AI-powered contract intelligence changes the day-to-day experience of working with contracts by making it possible to ask questions and get immediate answers.

The category manager who is about to have a conversation with a supplier about expanding scope can ask the AI:

"What does our current contract say about scope changes? Is there a change management provision? Does it specify a pricing methodology for out-of-scope work? Are there any limitations on scope expansion without a formal amendment?"

The procurement director who is briefing the CFO on supplier risk can ask:

"Across our top fifty suppliers by spend, how many have liability caps below $10 million? How many lack adequate

data breach notification provisions? What's our aggregate maximum exposure under the indemnification provisions?"

The contract negotiator who is about to enter a renewal discussion can ask:

"Pull every clause in our current agreement that deviates from the template we've been using for new agreements in this category. Rank the deviations by commercial impact and recommend which ones to prioritize in the renewal negotiation."

Each of these questions, in the pre-AI world, would have required someone to read through the relevant contracts, sometimes dozens of them, and manually extract and compile the answer. The delay alone often meant the question went unasked, because by the time someone could answer it, the decision had already been made based on incomplete information.

The ability to ask contract questions and get immediate, accurate answers changes how procurement professionals interact with their contracts. Contracts become a living resource that informs decisions in real time rather than a static archive that's consulted only when something goes wrong. That shift, from contract as document to contract as intelligence, is the core transformation that AI enables.

THE CONTRACT LIFECYCLE, REIMAGINED

Let me step back and paint a bigger picture of how AI transforms the entire contract lifecycle from initial drafting through active management to renewal or termination.

Pre-Signature: Smarter Drafting and Negotiation

Before a contract is signed, AI can assist with drafting, review, and negotiation in ways that produce better agreements from the outset.

Template optimization.

AI can analyze your existing contract templates against market benchmarks, identifying provisions that are weaker than industry standard and recommending improvements. It can also analyze the outcomes of past negotiations (which provisions were most frequently negotiated, which concessions were most commonly made) to help you build templates that are more likely to survive negotiation intact.

Redline analysis.

When a supplier returns your contract with their redlines, AI can instantly categorize each change by type and risk level.

- Cosmetic changes.
- Clarifications that don't alter the commercial or legal meaning.
- Substantive changes that shift risk or reduce your protections.
- And the critical changes that materially affect pricing, liability, indemnification, or termination rights that demand your most careful attention.

This triage function alone saves hours per contract and ensures that the most important changes get the most attention.

Clause benchmarking.

For any specific clause that's under negotiation, AI can show you how similar clauses are structured in comparable agreements across your portfolio and, if you're using a platform with community data, across the broader market.

"Is the supplier's proposed limitation of liability consistent with what we've accepted from other technology suppliers of similar size?"

The AI can answer that question in seconds, giving you evidence-based negotiation leverage rather than a gut-feel response.

Post-Signature: Active Contract Management

After the contract is signed, AI transforms contract management from passive document storage to active value management.

Performance monitoring.

AI can continuously compare operational performance data (delivery times, quality metrics, service levels) against the contractual definitions and thresholds. When performance dips below the contractual standard, the system flags it in real time rather than waiting for someone to notice during a quarterly business review. This early detection allows for prompt intervention (a conversation with the supplier, a formal notice under the contract's remediation provisions, or a commercial adjustment) before the performance gap compounds into a serious problem.

Invoice compliance.

AI can match invoices against contracted rates and terms automatically, flagging discrepancies before payment is processed. Is the supplier billing at the contracted rate? Are the volume discounts being applied correctly? Are the payment terms on the invoice consistent with what the contract specifies? This kind of invoice-to-contract matching has always been theoretically possible through manual audit, but the volume of invoices in most organizations makes comprehensive manual review impractical. AI makes it feasible—and the value recovery from catching billing discrepancies routinely exceeds the cost of the technology.

Amendment tracking.

Over the life of a long-term contract, amendments accumulate. Change orders. Scope modifications. Pricing adjustments. Extension letters. Each amendment modifies the original agreement, and keeping track of the cumulative effect of multiple amendments on the contract's commercial terms is a significant challenge. AI can maintain a current, consolidated view of the contract as amended—showing the procurement professional not just what the original terms were, but what the effective terms are after all modifications.

Pre-Renewal: Intelligence-Driven Renewal Preparation

The renewal period is where contract intelligence delivers its most concentrated value—and where it connects most directly to the negotiation preparation framework from Chapter Seven.

Six months before a renewal, AI-powered contract intelligence should produce a renewal preparation package that includes: a summary of the current terms and how they compare to market standards, a history of the contract's amendments and price changes over its life, a performance assessment based on continuous monitoring data, a risk assessment based on the supplier's current profile, and a list of recommended terms improvements for the renewal negotiation.

The prompt I use to generate the renewal brief:

"This contract expires in [X] days. The auto-renewal notification deadline is [Y] days. Based on your review of the current terms, prepare a renewal preparation brief that includes: (1) Key terms that should be improved in the renewal—where are we below market standard? (2) Provisions that are missing and should be added—most-favored-customer clause, benchmarking rights, enhanced termination flexibility, etc. (3) Pricing provisions—how has our pricing compared to market movement over the contract term? (4) Performance-related provisions—based on the SLA

history, what should be tightened? (5) The top three negotiation priorities, ranked by impact."

Appendix A contains the full version plus the auto-renewal trigger detection variant.

This package becomes the starting point for Phase Five of the Chapter Seven preparation framework—the contract analysis phase that identifies the specific terms to target in the renewal. But because the analysis is generated automatically from the contract intelligence platform, the category manager receives it as a ready-made input rather than having to build it from scratch. The preparation framework from Chapter Seven still applies in full—the contract analysis just becomes faster and more comprehensive because the heavy lifting has been done continuously rather than sprinted at renewal time.

This is the year three problem, solved. Not through heroic last-minute preparation, though that's better than nothing, but through continuous, AI-powered contract management that ensures you're always ready for the renewal conversation because you've been managing the contract actively throughout its life rather than filing it away and forgetting about it until the deadline looms.

THE TRAPS: WHERE CONTRACT INTELLIGENCE GOES WRONG

As with every AI application we've discussed, contract intelligence has limitations and failure modes that the best buyers need to understand.

The extraction accuracy trap.

AI-powered contract data extraction is reliable, but it's not perfect. Complex contract language, unusual formatting, poor document quality (especially scanned PDFs), and non-standard terminology can all cause extraction errors. The consequence of an extraction error in contract intelligence can be significant. If the AI misreads a liability cap or a termination notice period, and you rely on that misread in a negotiation or a business decision, the consequences are real.

The mitigation is straightforward: validate AI extractions against the source document for any provision you're going to rely on for a significant decision. The AI extracts and organizes. You verify before you act. This validation step is non-negotiable for high-stakes provisions—pricing terms, liability caps, termination rights, auto-renewal triggers. For lower-stakes provisions (reporting requirements, standard compliance language) a sampling-based validation approach is appropriate.

The false completeness trap.

An AI-powered contract database looks comprehensive and authoritative. But it's only as complete as the contracts that were fed into it. If your electronic contract repository is incomplete, the AI's analysis will have gaps that it doesn't flag. The most dangerous contract management failure is the contract you don't know about—the agreement signed by a business unit that never involved procurement, the side letter that modified the master agreement, the amendment that was executed but never filed.

AI can't analyze what it hasn't been given.

The "set it and forget it" trap.

Contract intelligence platforms need to be maintained. New contracts need to be ingested. Amendments need to be added. Supplier data needs to be updated. Extraction models need to be refined as you discover errors. **Organizations that treat contract intelligence as a one-time implementation rather than a continuous capability will find that the database becomes progressively less accurate and less complete over time**—eventually reaching the point where users stop trusting it, which defeats the entire purpose.

THE COMPETITIVE ADVANTAGE: WHAT CONTRACT INTELLIGENCE MAKES POSSIBLE

Let me close this chapter, and Part Two, by connecting the capabilities we've discussed back to the fundamental question: what does this mean for procurement outcomes?

Organizations with mature contract intelligence capabilities operate differently from those without them.

- Their category managers walk into every renewal negotiation knowing exactly what their current terms are, how those terms compare to market standards, and which specific provisions represent the greatest opportunity for improvement.

- Their risk managers have a portfolio-level view of contractual exposures that allows them to identify and address concentrations before they become problems.
- Their finance teams have visibility into contractual commitments and obligations that enables more accurate forecasting and more proactive cash flow management.
- Their legal teams can assess the organization's contractual risk profile across the entire supplier portfolio rather than reviewing contracts one at a time in response to individual issues.

And perhaps most importantly, the procurement function itself operates with a level of commercial intelligence that transforms its credibility with the rest of the organization. When the CFO asks "what's our exposure if this supplier fails?" and procurement can answer immediately, with specific contract provisions and financial implications —that's a different interaction than "let me check and get back to you."

When the business unit president asks "can we expand scope with this supplier?" and procurement can immediately surface the relevant contract provisions, flag the commercial implications, and recommend a negotiation approach—that's a strategic advisory conversation, not an administrative inquiry.

Contract intelligence doesn't replace the procurement professional's judgment. It gives that judgment a foundation of information that was previously inaccessible. It makes the invisible visible. It turns the filing cabinet into a strategic weapon.

And with that, Part Two is complete.

We've walked through five domains where AI is transforming procurement practice: spend analysis, market intelligence, supplier risk, negotiation preparation, and contract management.

In each domain, the same pattern has emerged.

AI provides capabilities (speed, scale, pattern recognition, synthesis across thousands of data points) that no human could match.

Humans provide capabilities (judgment, creativity, organizational context, ethical reasoning, the relationship intelligence that determines what to do with what AI reveals) that no algorithm can replicate.

AI reveals the landscape. Human judgment navigates it.

The combination produces outcomes that neither could achieve alone. That isn't a tagline—it's the operating model that the rest of this book is going to make real.

The capabilities I've described—spend visibility at machine speed, supplier intelligence in ninety minutes, continuous risk monitoring, AI-augmented negotiation preparation, contract intelligence at portfolio scale—already exist. But most procurement functions have not yet captured the value that these capabilities make possible, and the gap between "AI is available" and "AI is creating value at this organization" is the gap that Part Three is about.

How do you build a team that can leverage these capabilities?

How do you implement AI in a procurement function without the change management trainwreck that derails so many technology initiatives?

How do you govern the use of AI responsibly—managing the bias, the data security, the ethical questions, the tradeoffs that come with delegating decisions to algorithms?

And what does the future of this profession look like for those who are ready to lead it through this transition?

That's Part Three.

PART THREE

BUILDING THE AI-ENABLED ORGANIZATION

FROM OPERATING MODEL DESIGN TO ETHICS TO THE FUTURE OF THE PROFESSION.

CHAPTER 9
THE HUMAN + AI OPERATING MODEL

REDESIGNING HOW PROCUREMENT TEAMS WORK

There's a conversation happening in procurement offices around the world right now that sounds something like this: "We've seen what AI can do. We've piloted some tools. A few people on the team are using ChatGPT for research and it's clearly faster. But how do we actually change the way our team works? How do we move from a few individual procurement professionals experimenting with AI to an entire procurement function that operates differently?"

That conversation is the right one to be having. And the answer is more nuanced than "give everyone an AI subscription and tell them to figure it out"—which is, unfortunately, the approach a distressing number of organizations are taking.

The transition from a traditional procurement operating model to an AI-augmented one is not primarily a technology challenge. It's an organizational design challenge. It requires rethinking roles, redefining skill requirements, redesigning workflows, and rebuilding the expectations that the rest of the organization has about what procurement can deliver and how fast it can deliver it.

Get this wrong, and you end up with expensive tools that nobody uses, frustrated procurement teams who feel like AI was imposed on them rather than empowering them, and an organization that concludes "AI doesn't work for procurement" when the real problem was never the technology—it was the change management.

Let me walk you through what the AI-augmented procurement operating model actually looks like as a practical blueprint that I've seen work in real organizations with real teams and real constraints.

If you lead a procurement function, Part Three is where this book turns from "what's possible" to "how do we actually capture it." The capabilities described in Part Two—spend visibility at machine speed, supplier intelligence in ninety minutes, continuous risk monitoring, AI-augmented negotiation preparation, contract intelligence at portfolio scale—exist today. The reason most procurement functions have not yet captured the value those capabilities make possible is rarely technological. It's organizational. It's how teams are structured, how roles are defined, how workflows are designed, how change is led. The argument of this chapter is that the operating model is the leverage point—and the procurement leaders who get the operating model right will deliver outcomes that the leaders who treat AI as a tool deployment never will. Part Three is the playbook for getting it right.

THE ROLE TRANSFORMATION: WHAT CHANGES AND WHAT DOESN'T

The most important thing to understand about AI's impact on procurement roles is that it doesn't eliminate roles—it transforms them. The procurement professional of 2026 does the same fundamental work as the procurement professional of 2016: they source goods and services, manage supplier relationships, negotiate contracts, mitigate risks, and create value for their organizations. But how they do that work (the tools they use, the time allocation across activities, the depth and speed of their analysis) is substantially different.

Let me paint this concretely. Consider a typical category manager's time allocation before AI adoption versus after.

Before AI:

Roughly fifty to sixty percent of a category manager's time was spent on data gathering, analysis preparation, and report generation—the mechanics of building the analytical foundation for strategic work. Pulling spend data. Building supplier profiles. Reading contracts. Compiling market research. Formatting presentations. The remaining forty to fifty percent was spent on the work that actually creates value—developing sourcing strategies, engaging with stakeholders, negotiating with suppliers, managing relationships, solving problems.

After AI adoption:

The ratio flips. Data gathering, analysis, and report preparation, now largely AI-assisted, consumes fifteen to twenty percent of the category manager's time. The remaining eighty to eighty-five percent is available for strategy, stakeholder engagement, negotiation, and relationship management. The same person, doing the same job, is now spending the vast majority of their time on the highest-value activities rather than the preparatory work that used to consume most of their bandwidth.

This isn't a theoretical projection. I've watched this transformation happen with real procurement professionals. The category manager I mentioned in Chapter Three—the one who said she was "finally doing the job I was hired to do instead of the data entry I was actually doing"—is representative of a pattern I've seen repeatedly. The procurement pros who adopt AI tools effectively don't work fewer hours. They work differently—covering more categories with greater depth, having more stakeholder conversations, preparing more thoroughly for negotiations, thinking more creatively about sourcing strategies. They do more of the work that creates value, because AI handles more of the work that used to consume their time.

The New Skill Profile

The transformation in how procurement professionals spend their time implies a transformation in the skills they need. The skill profile for procurement excellence is shifting—not away from the fundamentals we discussed in Part One, but toward a combination of traditional procurement expertise and new capabilities that amplify that expertise through technology.

Skills that become more important:

Strategic thinking and business acumen.

When AI handles the analytical mechanics, the ability to interpret data, develop creative strategies, and navigate organizational complexity increases substantially. The category manager who can analyze but not strategize was always underperforming. In the AI era, the gap between analytical capability and strategic capability becomes the defining differentiator.

Stakeholder management and influence.

As procurement's analytical capabilities expand through AI, the opportunity to engage with the business at a more strategic level grows correspondingly. The sourcing professionals who can translate AI-powered intelligence into compelling business conversations, who can brief a divisional VP with the kind of data-backed insight we discussed in Chapter Five, become significantly more valuable to their organizations.

Prompt engineering and AI literacy.

The ability to ask the right questions, structure the

right prompts, and critically evaluate the AI's output is a new core competency that didn't exist five years ago. It's not a technical skill in the traditional sense. As we discussed in Chapter Three, it's a communication skill. But it's a communication skill that requires deliberate development and practice.

Critical thinking about AI output.

The ability to evaluate whether an AI-generated analysis is accurate, complete, and relevant (to catch hallucinations, identify biases, recognize when the AI has made assumptions that don't fit your context) is essential. This is the "brilliant intern supervision" capability from Chapter Three, and it requires both domain expertise (to know when something doesn't look right) and AI literacy (to understand why the AI might have gone wrong).

Data storytelling.

The ability to translate complex analytical output into narratives that drive action, to take the AI's market analysis and turn it into a stakeholder presentation that generates support for a sourcing initiative, becomes a premium skill when the constraint is no longer generating analysis but communicating it effectively.

Skills that become less critical as standalone capabilities:

Manual data analysis and spreadsheet manipulation. These skills don't disappear entirely, there will always be situations where hands-on data work is necessary, but the premium on them as standalone capabilities declines as AI handles more of the analytical heavy lifting.

Report generation and formatting. The hours spent formatting PowerPoint presentations and building Excel dashboards are among the most directly displaced by AI tools that can generate polished output from analytical inputs.

Basic research and information gathering. The procurement analyst whose primary value was collecting and organizing information from various sources finds that value proposition challenged by AI tools that can do the same work faster and more comprehensively.

This skill shift has important implications for hiring, development, and team composition. But the fundamental point is this: the AI-augmented procurement professional is not a technology specialist who happens to work in procurement. They are a procurement specialist who has added AI literacy to their professional toolkit. The domain expertise (the procurement knowledge, the market understanding, the negotiation skill, the relationship intelligence) remains the foundation. AI is the amplifier.

THE BEFORE AND AFTER: WHAT THE SHIFT ACTUALLY LOOKS LIKE IN PRACTICE

Let me make the operating model transformation tangible with detailed before-and-after scenarios across three common procurement activities, because abstract organizational design means nothing if the best buyers can't see themselves in it.

Scenario One: The Category Strategy Refresh

Before AI:

The category manager for professional services is asked to refresh the sourcing strategy for consulting spend—approximately $15 million across four business units. She starts by pulling spend data, which requires coordinating with finance to get extracts from two ERP instances and a purchasing card platform. The data arrives in three different formats over the course of a week. She spends another week cleaning, normalizing, and classifying the data, discovering in the process that roughly twenty percent of consulting spend has been miscoded to "other professional services" and needs to be reclassified manually by reading invoice descriptions one by one.

With cleaned spend data in hand, she begins market research. She reads two analyst reports on the consulting market, calls three colleagues at other companies to get informal benchmarks on rates, and reviews the incumbent suppliers' websites and recent press releases. This takes another week and a half. She then builds a PowerPoint deck summarizing her findings and recommendations—three days of formatting, chart-building, and wordsmithing to make it presentable for the leadership review. Total elapsed time from assignment to presentation: approximately six weeks. During those six weeks, she has done almost no strategic thinking—she has been collecting, cleaning, organizing, and formatting information. The strategic analysis that should have been the point of the exercise gets compressed into whatever time is left, which is usually a few hours the day before the presentation.

After AI:

Same category manager, same assignment. She starts by uploading the raw spend data—all three sources, all three formats—into an AI tool that classifies, normalizes, and consolidates the data in about four hours. She spot-checks the classification, corrects a handful of errors, and has a clean spend picture by the end of day one.

Day two, she uses the Chapter Five market intelligence workflow to build a comprehensive consulting market analysis—competitive landscape, pricing trends, emerging delivery models, labor market dynamics. This takes about three hours and produces a more thorough market assessment than the manual research would have generated in ten days, because the AI synthesizes information from dozens of sources rather than the handful she would have time to read manually.

Days three and four, she does the work that matters: strategic analysis. Examining the spend patterns against the market intelligence, she identifies that the organization is paying an average of $285 per hour for capabilities that the market prices at $230-260. Two of the four business units, she discovers, are using the same consulting firm for similar work but negotiated independently at different rates—one paying $310 per hour and the other $255 for substantially identical services. And there are three qualified alternative providers that entered the market in the past two years that none of the incumbent relationships have been competitively tested against.

She develops a sourcing strategy with three options: full

competitive rebid, incumbent renegotiation with competitive pressure, and a hybrid approach that consolidates volume with two preferred providers. The expected savings for each option get modeled out, and a stakeholder engagement plan for each scenario identifies which business unit leaders need to be brought on board and in what sequence.

Day five, she presents to leadership. The presentation took her about ninety minutes to build, not three days, because the AI generated the charts, formatted the data, and drafted the executive summary. She spent her time reviewing and refining rather than producing from scratch.

Total elapsed time: five business days. Not six weeks.

Five days. And the quality of the strategic analysis is significantly higher because she spent eighty percent of those five days thinking rather than collecting and formatting.

Scenario Two: The Supplier Risk Response

Before AI:

A Friday afternoon call from the VP of Manufacturing: "I just heard that one of our packaging suppliers may have been bought by a Chinese company. Is that going to affect our supply? Are there regulatory implications? I need answers by Monday."

The category manager spends Friday evening and Saturday morning trying to track down the rumor. The supplier's account manager is vague and says they "can't comment on M&A activity." News sources turn up a brief mention in a trade publication but no details. She tries to assess the regulatory implications but isn't sure which regulations might apply to Chinese ownership of a packaging supplier. Sunday is spent building a PowerPoint that mostly says "we don't know enough yet" with some generic risk mitigation recommendations.

The Monday morning meeting is unsatisfying for everyone—the VP wants certainty she can't provide, the risk assessment is thin, the mitigation plan is generic. Everyone agrees to "monitor the situation"—which, in practice, means waiting until something happens and then reacting.

After AI:

Same Friday afternoon call. The category manager opens her AI tool and begins a focused intelligence sprint. She asks the AI to search for any M&A activity involving the supplier, any regulatory filings related to Chinese acquisition of U.S. packaging companies, and any recent CFIUS (Committee on Foreign Investment in the United States) reviews in the packaging industry. Within thirty minutes, she has a comprehensive picture: the acquisition is real, announced in an overseas financial publication three days ago, and involves a Chinese materials conglomerate that has previously acquired two other packaging companies in Europe.

She then asks the AI to assess the regulatory implications: does CFIUS review apply? (Probably not for packaging, unless the supplier serves defense or critical infrastructure clients.) Are there any sanctions or trade restriction implications? (Detailed analysis of current regulations.) What are the precedents for similar acquisitions? (Three comparable cases identified with their outcomes.)

She asks the AI to pull the change-of-control provisions from her current contract with the supplier: does the acquisition trigger any contractual rights? (Yes—there's a termination-for-convenience clause that's triggered by a change of control, giving her organization sixty days to exit if they choose.) She asks for a quick assessment of alternative suppliers who could absorb the volume if a transition is needed.

By Saturday morning, less than eighteen hours after the call, she has a comprehensive briefing that includes: confirmed details of the acquisition, regulatory risk assessment with precedent analysis, contractual implications including the change-of-control provision, three alternative suppliers with capability assessments, and a recommended action plan with three scenarios (stay and monitor, negotiate enhanced protections, begin transition planning).

The Monday meeting is substantive and productive. The VP has the information she needs to make a decision. The team has a clear action plan. And the category manager has demonstrated a level of responsiveness and analytical depth that changes how the executive team perceives procurement's capability.

Scenario Three: The Routine Negotiation That Became Strategic

Before AI:

A $1.2 million annual contract for temporary staffing is up for renewal. In the old model, this is a tier-two negotiation that gets maybe a day of preparation. The category manager reviews last year's pricing, has a brief conversation with the hiring managers who use the service, and walks into the renewal meeting with a general sense that "we should push for flat pricing." The supplier pushes for a five percent increase citing labor market tightness. They settle on three percent. Both parties feel the outcome was reasonable. Nobody explores whether the outcome was optimal.

After AI:

Same contract, same dollar value. But now the category manager runs it through an abbreviated version of the Chapter Seven preparation framework —about two hours of AI-powered research. The market analysis reveals that bill rates in this geography for the skill sets they're buying have actually declined two percent year-over-year. The supplier financial analysis shows the company just reported record revenue and expanding margins. The contract analysis reveals that the agreement lacks a most-favored-customer clause and has a termination notice period that's thirty days longer than market standard. And a competitive scan identifies two alternative providers with strong reviews from companies in similar industries.

Armed with this preparation, the negotiation produces a different result: flat pricing (justified by the declining rate market), a most-favored-customer clause added to the contract, the termination notice period reduced from ninety to sixty days, and improved service level metrics tied to candidate quality. The total incremental value versus the "push for flat pricing" approach: approximately $180,000 over the two-year term, plus structural improvements in the contract terms that create value beyond the immediate pricing.

Two hours of AI-powered preparation. $180,000 in incremental value on a contract that "wasn't big enough" to justify thorough preparation in the old model. Multiply that across the thirty or forty mid-tier negotiations in a typical procurement portfolio, and the portfolio-level impact becomes transformative.

TEAM DESIGN: STRUCTURING THE AI-AUGMENTED PROCUREMENT FUNCTION

How should a procurement organization be structured to maximize the value of AI capabilities? The answer depends on the organization's size, maturity, and technology investment level, but some principles apply broadly.

The Hub-and-Spoke Model

The model I've seen work most effectively is what I call the hub-and-spoke approach. A small central team, the hub, develops AI capabilities, tools, and workflows and supports the broader procurement organization in adopting them. The category managers and sourcing professionals, the spokes, apply those capabilities in their day-to-day work, bringing domain expertise and organizational context that the central team can't replicate.

The hub team typically includes a procurement AI lead —someone who combines procurement expertise with strong AI literacy and who serves as the internal champion for AI adoption. It might include a data analyst or data engineer who manages the technical infrastructure—data pipelines, platform integrations, model configuration. And it might include a change management or training resource who helps procurement professionals develop the skills to use AI tools effectively.

The hub team's job is not to do AI-powered analysis for the category managers. It's to build the capabilities, tools, and training that enable category managers to do it themselves. This is a critical distinction. If the hub becomes a bottleneck, if category managers have to submit requests to the AI team and wait for results, you've just recreated the old model with fancier technology. The goal is to democratize the capability so that every procurement professional can leverage AI independently in their daily work.

The Center of Excellence Evolution

Many procurement organizations already have a sourcing center of excellence—a team that develops sourcing methodologies, maintains templates and tools, provides training, and supports complex sourcing events. The natural evolution is to expand this center of excellence to incorporate AI capabilities—adding AI-powered workflows to the sourcing methodology, building prompt libraries for common procurement tasks, developing training curricula that include AI literacy, and establishing governance frameworks for AI use in procurement.

This evolution works well because it builds on existing

organizational structures and relationships rather than requiring new ones. The center of excellence already has credibility with the procurement team. It already has mechanisms for developing and deploying best practices. Adding AI to its mandate is an expansion, not a reinvention.

The Embedded Specialist Model

For larger procurement organizations, a third model involves embedding AI-skilled procurement teams within category teams—individuals who combine procurement knowledge with advanced AI capabilities and who work alongside category managers on a daily basis. These embedded specialists are not data scientists. They're procurement professionals who have developed deeper AI skills than the average procurement pro and who can handle more complex AI applications (building custom analytical workflows, configuring specialized tools, developing category-specific prompt libraries) that are beyond what the typical category manager needs to do.

The embedded specialist model is particularly effective for large, complex categories where the AI applications are sophisticated enough to warrant dedicated expertise. The technology category manager negotiating a nine-figure cloud platform contract benefits from having someone on the team who can build a comprehensive AI-powered supplier intelligence platform for that specific negotiation —something more complex and customized than the general-purpose workflows we discussed in Chapter Seven.

Most of these implementations don't fail because the

technology was wrong or the strategy was bad. They fail because somebody underestimated how hard it is to get people to change how they work. Plan for the change management problem the same way you plan for the technology problem. Probably more, actually.

MANAGING THE TRANSITION: FROM RESISTANCE TO ADOPTION

Let me be honest about something that the AI enthusiasm often glosses over: most procurement professionals are not eager to adopt AI. Not because they're Luddites or technophobes, though some are, but because they're busy, they're skeptical of technology that promises to transform their work (they've heard that before, from ERP vendors and e-sourcing platform sellers and every other technology wave that promised revolution and delivered incremental improvement), and they're understandably nervous about the implication that AI might make their skills obsolete.

Managing this resistance is not optional. It is a core leadership challenge that will determine whether your AI investment produces value or becomes another shelfware story.

Here's what I've seen work.

Start with a real pain point, not a technology demonstration.

Don't begin your AI rollout by showing the team what the technology can do. Begin by asking them what's making their work harder, slower, or more frustrating than it needs to be.

The category manager who says "I spend two weeks building a spend analysis every quarter" is telling you exactly where to start.

Show them AI-powered spend classification that cuts that timeline to two days, and you have an advocate rather than a skeptic. The sourcing professional who discovers that AI solves a problem they actually care about becomes your most effective internal champion—far more persuasive than any management directive.

Make the first experience a win.

The first time a procurement professional uses an AI tool should produce a result that they couldn't have produced as quickly, an insight they wouldn't have found, a draft that's good enough to be useful. If the first experience is a hallucination, a generic response, or a technically impressive but practically useless output, you've lost that procurement professional for months.

Invest the effort to ensure that early experiences are positive by providing guidance on effective prompts, selecting use cases where AI is most reliably useful, and pairing new users with experienced ones who can help them avoid common mistakes.

Don't mandate—enable.

The procurement leaders who try to force AI adoption through mandates, "every sourcing event must include AI-powered market intelligence", consistently generate more resistance than adoption. The leaders who enable adoption (providing tools, training, time, and support) and then let the results speak for themselves consistently generate more sustainable change. When procurement professionals see their colleagues preparing for negotiations in ninety minutes instead of two weeks, they want to learn. That pull is more powerful than any push.

Acknowledge the fear honestly.

The "will AI take my job?" question from Chapter Three doesn't go away when you start implementation. Address it directly, honestly, and repeatedly. Show procurement pros specifically how AI changes their work rather than eliminates it. Share examples of colleagues who have adopted AI and are doing more interesting, more strategic, more impactful work as a result. Be honest about the tasks that are being automated—and be equally clear about the tasks that are becoming more important and more valued.

Invest in training that's practical, not theoretical.

Nobody needs a lecture on machine learning architectures. What sourcing professionals need is hands-on training that shows them exactly how to use AI tools for the procurement tasks they perform every day.—how to build a supplier dossier, how to analyze a contract, how to prepare for a negotiation, how to run a spend classification.

Provide training that produces an output they can actually use (an analysis for a real negotiation they're preparing for, a contract review for a real renewal they're managing) is training that sticks. Abstract training is forgotten by the next morning.

MEASURING SUCCESS: THE METRICS THAT MATTER

How do you know if your AI-augmented operating model is working? The temptation is to measure AI adoption itself—how many buyers are using the tools, how many prompts they're running, how many analyses they're generating. Resist that temptation. Activity metrics measure motion, not value.

The metrics that matter are procurement outcome metrics that improve because of AI adoption.

Time to insight.

How long does it take from identifying a sourcing opportunity to having the intelligence required to develop a strategy? In the pre-AI model, this was typically measured in weeks. In the AI-augmented model, it should be measured in days or hours. If this metric isn't improving, your procurement professionals either aren't using the tools or aren't using them effectively.

Negotiation outcomes.

Are negotiation results improving—not just on price, but on total value including terms, service levels, risk protections, and relationship quality? The preparation framework from Chapter Seven should produce measurably better outcomes across the negotiation portfolio. Track it.

Contract value realization.

Are you capturing more of the value that's negotiated into your contracts—fewer missed auto-renewals, fewer pricing errors going undetected, more contractual entitlements being exercised? The contract intelligence capabilities from Chapter Eight should show up in this metric.

Portfolio coverage.

How many of your spend categories have current, evidence-based sourcing strategies? In the pre-AI model, comprehensive strategy coverage was limited by research bandwidth. In the AI-augmented model, it should expand significantly. If you're still only strategically sourcing your top ten categories because "we don't have time for the rest," your AI adoption isn't translating into the portfolio-level impact it should.

Stakeholder satisfaction.

Are business stakeholders experiencing procurement as more strategic, more responsive, and more value-creating? This is the ultimate measure of whether the operating model transformation is working—not in procurement's self-assessment, but in the eyes of the business partners who experience the results.

Procurement pro satisfaction.

Are your procurement professionals finding their work more interesting, more strategic, and more professionally fulfilling? The procurement teams who are spending eighty percent of their time on strategy and relationships instead of sixty percent on data entry should be more engaged. If they're not, if the AI adoption is creating stress rather than reducing it, something in the implementation needs adjustment.

THE LEADERSHIP IMPERATIVE

I want to take a moment for a direct message to procurement leaders (CPOs, VPs, directors) because the success or failure of the AI-augmented operating model depends more on leadership than on any other single factor.

Your team is watching you. If you personally use AI tools in your work (if you reference AI-generated intelligence in your executive presentations, if you use AI to prepare for your own negotiations, if you visibly and consistently demonstrate that AI is part of how you work) your team will follow. If you delegate AI to the innovation team, mention it in town halls but never use it yourself, and continue operating the way you always have—your team will conclude that AI is a corporate initiative to be endured rather than a capability to be embraced.

Lead by example. Use the tools. Share what you learn. Be honest about what works and what doesn't. Show your team that adopting AI is not a threat to their professional identity but an evolution of it—an evolution that makes the work more interesting, more impactful, and more valued.

This leadership posture isn't a nice-to-have. The procurement leaders who personally model AI adoption, build AI-augmented teams, and design workflows that maximize the technology's value are going to deliver results that redefine what their organizations expect from procurement. The leaders who don't will find themselves at a structural disadvantage that compounds with every quarter.

THE GENERATIONAL DYNAMIC: BRIDGING EXPERIENCE AND DIGITAL FLUENCY

One of the most important and most underappreciated challenges in transitioning to an AI-augmented operating model is managing the generational dynamic within procurement teams. There are two distinct populations in most procurement organizations, and they bring different strengths and different anxieties to the AI transition.

On one side, you have the experienced procurement professionals with twenty or thirty years of procurement experience, deep domain expertise, and well-developed instincts about markets, suppliers, and negotiation. They are often more skeptical of AI and slower to adopt it, partly because they're confident in their existing capabilities and partly because the hands-on, learn-by-doing pattern that built their expertise doesn't translate easily to a tool that requires comfort with technology and trust in unfamiliar processes.

On the other side, you have the younger procurement pros with five to ten years of experience, digital natives who grew up with technology and adopt new tools as naturally as breathing. They are enthusiastic about AI, quick to experiment, and often produce impressive AI-powered analysis within days of getting access to the tools. What they lack is the domain expertise that makes AI truly powerful—the thirty years of market knowledge, supplier relationship capital, and organizational navigation skill that their senior colleagues possess.

The organizations that manage this tension well are the ones that frame it explicitly as complementary rather than competitive. The senior sourcing professional doesn't need to become an AI expert. They need to learn enough to direct AI tools effectively—which, as we discussed in Chapter Three, is a communication skill that leverages exactly the kind of expertise they already have. The junior procurement professional doesn't need thirty years of experience to add value. They need to pair their digital fluency with deliberate mentorship from colleagues who can teach them what questions to ask and how to evaluate whether the AI's answers actually make sense in a procurement context.

I've seen this pairing work beautifully in practice. At one organization, they created what they called "AI tandems"—deliberately pairing a senior category manager with a junior analyst for each major sourcing initiative. The senior procurement professional directed the strategy and defined the intelligence requirements. The junior procurement pro executed the AI-powered research and analysis workflows. Together, they produced work that was both strategically sophisticated and analytically comprehensive—better than either could have produced alone.

The senior sourcing professional in one of these tandems told me something that stuck with me:

"I've been doing this for twenty-five years, and I've never been as well-prepared for a negotiation as I was when Sarah built the dossier for our packaging renewal. She did in three hours what would have taken me two weeks. But she couldn't have done it without my telling her what to research and why it mattered."

That's the model—experience directing capability, domain expertise amplifying analytical power. Two generations of sourcing professionals finding that they need each other more, not less, in the AI era.

The leaders who foster this dynamic (who create structures, incentives, and cultural norms that encourage cross-generational collaboration rather than competition) are going to have significantly more effective teams than the leaders who let the generational tension fester into resentment on both sides.

THE TRAINING CURRICULUM: WHAT TO TEACH AND HOW TO TEACH IT

Let me get specific about training, because "invest in training" is advice that everyone gives and nobody makes practical. Here's the training curriculum I've developed through my teaching at Rutgers and my work with corporate procurement teams, organized into three tiers.

Tier One: AI Literacy for All Buyers (One Day)

Every procurement professional—regardless of role, seniority, or technical comfort level—needs a baseline understanding of what AI is, what it can do, what it can't do, and how to use it responsibly. This is the Chapter Three content, delivered in a practical, hands-on format.

The morning session covers the fundamentals: what machine learning, NLP, and generative AI actually are, explained through procurement-specific examples rather than abstract technology concepts. The hallucination problem, with live demonstrations of AI getting things wrong so procurement professionals see it firsthand rather than just hearing about it. The "brilliant intern" model for how to think about AI output. And the data security and ethical guidelines that govern AI use in the organization.

The afternoon session is entirely hands-on. Every participant brings a real procurement task (a negotiation they're preparing for, a category they're researching, a contract they need to review) and works through it using an AI tool with live coaching from experienced procurement teams. The goal is that every participant leaves with a completed piece of work that they produced using AI and that they can use in their actual job. Not a theoretical exercise. A real deliverable.

This is non-negotiable training. Everyone gets it. It's the foundation everything else builds on.

Tier Two: Functional Skills for Category Managers and Sourcing Professionals (Two Days)

The second tier is targeted at the procurement professionals who will use AI most intensively in their daily work—category managers, sourcing professionals, contract managers, and procurement analysts. It covers the workflows from Part Two of this book: spend analysis, market intelligence, supplier risk monitoring, negotiation preparation, and contract analysis.

Day one covers spend analysis and market intelligence—the Chapter Four and Chapter Five workflows. Procurement pros work through the complete spend classification workflow using their organization's actual data. They build supplier dossiers for real upcoming negotiations. They construct market landscape analyses for their assigned categories. Every exercise uses real data and produces real deliverables.

Day two covers negotiation preparation and contract intelligence—the Chapter Seven and Chapter Eight workflows. Sourcing professionals prepare for an actual upcoming negotiation using the full seven-phase framework. They analyze real contracts from their portfolio. They war-game a negotiation scenario with the AI playing the supplier role. Again, every exercise produces output they can actually use.

The critical design principle is specificity over generality. Don't teach "how to use AI for procurement." Teach "how to build a supplier financial dossier for your upcoming renewal with [specific supplier name] using [specific AI tool]." The more specific and concrete the training, the more likely the best buyers are to apply it after they leave the room.

Tier Three: Advanced Applications for AI Champions (Ongoing)

The third tier is for the procurement professionals who emerge as natural AI champions—the ones who adopt quickly, experiment enthusiastically, and start finding applications that nobody anticipated. These procurement teams need advanced training in prompt engineering, custom workflow development, and the evaluation of purpose-built procurement AI platforms.

This tier is less structured and more organic. It might involve a monthly community of practice where AI champions share what they've learned, troubleshoot challenges, and collaborate on developing new applications. It might involve sending champions to AI-focused conferences or training programs. It might involve partnering with the organization's IT team or data science group on more sophisticated AI projects.

The investment in this tier is relatively small in terms of formal training time, but the return is disproportionately large. The AI champions become the organization's internal consultants, helping their colleagues adopt tools and workflows, solving problems that arise in practice, and continuously pushing the frontier of what AI-augmented procurement looks like in their specific organizational context.

THE WORKFLOW REDESIGN: MAKING AI NATIVE, NOT BOLTED ON

One of the most common mistakes in AI adoption is treating AI tools as additions to existing workflows rather than catalysts for redesigning workflows. The procurement team that adds "check with AI" as a step in their existing sourcing methodology has bolted AI onto an old process. The procurement team that has redesigned their sourcing methodology around AI capabilities has built something different and more effective.

Let me illustrate with a specific example: the strategic sourcing process.

The traditional seven-step strategic sourcing process (needs assessment, spend analysis, market research, strategy development, supplier selection, negotiation, and implementation) was designed for a world where each step required weeks of manual effort. The timelines, the gate reviews, and the stakeholder checkpoints were all calibrated to a pace of work that assumed two weeks for spend analysis, three weeks for market research, and similar timeframes for each subsequent step. A complete strategic sourcing cycle might take four to six months from initiation to contract signature.

In the AI-augmented model, the analytical steps compress significantly. Spend analysis that took two weeks takes two days. Market research that took three weeks takes three hours. Supplier evaluation that required two weeks of reference calls and proposal analysis takes a few days. The total cycle time can compress from six months to six to eight weeks without sacrificing analytical rigor because the AI is doing the heavy analytical lifting faster and more comprehensively than the manual process ever could.

But here's the critical insight: compressing the analytical timeline doesn't mean you should compress the strategic timeline proportionally.

The time you save on analysis should be reinvested in the phases that create the most value: deeper stakeholder engagement, more creative strategy development, more thorough negotiation preparation, and more careful supplier transition planning.

The goal isn't to do the same process faster. It's to do a better process in less time, devoting more attention to the judgment-intensive, relationship-dependent phases that AI can't handle and less time to the analytical phases that AI handles well.

This workflow redesign requires deliberate effort. Left to their own devices, most procurement teams will simply do the same process faster and then wonder why the outcomes haven't improved as much as expected. The improvement comes from redirecting the saved time toward higher-value activities—and that redirection needs to be designed into the workflow, not left to chance.

I worked with a procurement team that redesigned their sourcing methodology explicitly around AI capabilities. They compressed their analytical phases from twelve weeks to three weeks. But instead of delivering the sourcing outcome nine weeks earlier, they used six of those nine saved weeks to deepen their stakeholder engagement by having more conversations with more business users, understanding requirements more thoroughly, and building broader organizational support for the sourcing strategy. The remaining three weeks of time savings showed up as faster time-to-value.

The result? Their sourcing initiatives started delivering measurably better outcomes—not because the AI produced better analysis (though it did), but because the freed-up time allowed for the deeper stakeholder work that had always been skipped or rushed in the old process. The AI improved the analysis. The workflow redesign improved the strategy. Together, they transformed the sourcing function's effectiveness.

BUILDING CROSS-FUNCTIONAL AI PARTNERSHIPS

The AI-augmented procurement function doesn't operate in isolation. It intersects with, and increasingly depends on, partnerships with other functions that are also navigating their own AI transformations. The procurement leaders who get the operating model right will have built deliberate working relationships with three functions in particular.

The Procurement-Finance Partnership

Finance teams are among the most advanced AI adopters in most organizations, and the intersection between procurement AI and finance AI creates powerful synergies. When procurement's spend analytics platform is integrated with finance's budgeting and forecasting systems, both functions gain capabilities neither has alone. Finance gets real-time visibility into procurement commitments and spending trends that improve forecast accuracy. Procurement gets budget and financial planning data that informs sourcing strategy and timing.

The practical collaboration often starts with data.

Procurement and finance typically maintain separate views of the same spending—procurement sees it by supplier and category, finance sees it by cost center and GL code. AI-powered integration can reconcile these views, creating a unified spending picture that serves both functions. The procurement leader who proactively builds this bridge with their CFO counterpart creates an ally who understands and supports procurement's AI investments because they're benefiting from them directly.

The Procurement-Legal Partnership

Contract intelligence—the capabilities we discussed in Chapter Eight—sits squarely at the intersection of procurement and legal. The most effective implementations involve both functions as co-owners of the contract intelligence platform, with procurement driving the commercial analysis and legal driving the risk and compliance analysis.

I've seen organizations where procurement implemented contract intelligence in isolation, without legal involvement. The tools extracted commercial terms effectively but missed legal nuances (unusual indemnification structures, governing law implications, regulatory compliance gaps) that a legal-informed implementation would have caught. The reverse is equally suboptimal: legal teams that implement contract intelligence focused entirely on risk and compliance but miss the commercial optimization opportunities that procurement would have identified.

The partnership model works better than either function operating alone. Procurement and legal jointly define the extraction taxonomy—what terms matter and why. They jointly validate the AI's extractions—procurement checking commercial terms, legal checking risk provisions. And they jointly act on the intelligence—procurement driving commercial optimization, legal driving risk remediation, both functions informed by the same comprehensive view of the contract portfolio.

The Procurement-IT Partnership

The IT-procurement relationship has historically been transactional—IT defines requirements, procurement runs the sourcing, both sides occasionally pretend the other doesn't exist. AI changes the dynamic substantially. There's a substantive capability partnership that needs to be built. IT provides the infrastructure, security governance, and integration capabilities that procurement AI depends on. Procurement provides the use case definition, workflow design, and outcome accountability that gives IT's infrastructure investment a clear business purpose.

The most effective procurement-IT partnerships I've seen share several characteristics. There's a clear division of responsibility—IT owns the infrastructure and security, procurement owns the use cases and outcomes. There's regular communication—not just when something breaks, but proactive alignment on roadmaps, priorities, and resource needs. And there's mutual respect—IT doesn't dismiss procurement's requirements as "just another business unit wanting a shiny tool," and procurement doesn't dismiss IT's security and governance concerns as "bureaucratic obstacles."

The procurement leader who invests in building these cross-functional partnerships (finance, legal, IT) is not just building better AI capability. They're building organizational influence that extends procurement's strategic reach and ensures that procurement's AI investments are reinforced rather than undermined by adjacent functions.

THE CHANGE MEASUREMENT FRAMEWORK: TRACKING WHAT MATTERS OVER TIME

Let me expand on the metrics discussion with a practical framework for measuring the impact of your AI-augmented operating model over time. This framework tracks three time horizons.

Short-Term Metrics (First 90 Days)

In the first ninety days, you're measuring adoption and initial impact. How many procurement professionals have completed Tier One training? How many are actively using

AI tools at least weekly? What's the time savings on workflows—spend analysis cycle time, negotiation preparation time, contract review time? Has the team produced any measurable wins (better negotiation outcomes, identified savings opportunities, caught contract risks) that can be directly attributed to AI-assisted preparation?

These metrics establish the baseline and generate the early success stories that fuel broader adoption. Don't overcomplicate them. Simple counts and time comparisons are sufficient at this stage.

Medium-Term Metrics (Months 4-12)

By the four-to-twelve month mark, you should be measuring outcome improvements across the portfolio. Are negotiation outcomes improving on average? Is contract compliance increasing? Is spend under management expanding? Are more categories receiving current, evidence-based sourcing strategies? Are stakeholder satisfaction scores improving?

These metrics tell you whether AI adoption is translating into procurement performance improvement—which is the only thing that ultimately matters. If procurement pros are using AI tools actively but procurement outcomes aren't improving, something in the workflow redesign or the application of AI output needs adjustment.

Long-Term Metrics (Year Two and Beyond)

In the second year and beyond, you should be measuring strategic impact.

Has procurement's organizational positioning changed —is the function being included in more strategic conversations, consulted earlier in business decisions, valued more highly by executive leadership?

Has the procurement team's capacity expanded—are they covering more categories, managing more complex initiatives, delivering more strategic value with the same or similar headcount?

Has talent attraction and retention improved—are better candidates applying for procurement roles, and are your best sourcing professionals staying rather than leaving for more forward-thinking organizations?

These are the metrics that tell you whether the AI-augmented operating model is creating the kind of sustained competitive advantage that justifies ongoing investment and continuous evolution. They're harder to measure than cycle times and negotiation savings, but they're ultimately more important—because they reflect whether AI has transformed procurement's role in the organization or merely made it more efficient.

This is the operating model. Roles transformed, time allocation flipped, teams structured for AI-augmented work, transitions managed deliberately, training calibrated to the best buyers' actual jobs, workflows redesigned rather than bolted onto, partnerships built across the functions procurement depends on, and outcomes measured with discipline at every horizon. None of it is theoretical. All of it has been built and proven at organizations that decided to take this transition seriously.

The choice is yours. But I'd recommend choosing quickly, because the leaders who started six months ago already have a head start that is compounding with every negotiation, every sourcing event, and every contract renewal.

CHAPTER 10
IMPLEMENTATION

FROM PILOT TO SCALE WITHOUT THE TRAINWRECK

Every procurement technology implementation I've ever been involved with—and I've been involved with more than I can count across four Fortune 500 companies—has followed one of two trajectories. Either it started small, proved value quickly, built momentum through demonstrated results, and scaled organically because procurement professionals wanted more of what was working. Or it started with a grand vision, a massive budget, a twelve-month implementation timeline, a consulting firm's beautifully formatted transformation roadmap, and ended in a spectacular flameout that set the organization's technology adoption back by years and made the word "transformation" radioactive in procurement leadership meetings for the foreseeable future.

The difference between these two trajectories is not the

technology. The technology, in both cases, was capable of delivering value. The difference is the implementation approach—and specifically, whether the implementation was designed to produce early, visible, undeniable wins that created organizational momentum, or whether it was designed to achieve theoretical perfection before anyone saw a result.

I am passionately, vehemently, almost religiously committed to the first approach. Not because I'm impatient, though I am, but because I've watched too many procurement organizations invest millions of dollars and years of effort in technology implementations that produced beautiful architectures, comprehensive data models, and fully integrated platforms that nobody actually used because by the time the implementation was "complete," the procurement teams had lost faith, the executive sponsors had moved on, and the organizational energy required to drive adoption had long since dissipated.

AI implementation in procurement should be fast, focused, and relentlessly value-oriented. Let me show you how.

A note for senior leaders before we go further:

Of all the chapters in this book, this is the one most directly built for action. Chapter Nine described what an AI-augmented procurement operating model looks like; this chapter is the playbook for getting there from where you are today. The argument is straightforward: implementation success in procurement AI does not depend on selecting the right platform, securing the largest budget, or hiring the most expensive consultants. It depends on starting with one procurement professional, one negotiation, one quick win—and scaling from proven value rather than projected value. The procurement leaders who internalize this discipline produce sustained results. The leaders who default to the comprehensive transformation playbook produce the spectacular flameouts I just described. The chapter that follows is structured to help you stay in the first category.

THE NINETY-DAY QUICK WIN: WHERE TO START

If you take one thing from this chapter, let it be this:

Your AI implementation should produce its first measurable, communicable win within ninety days.

Not ninety days of planning. Ninety days from the moment you decide to start. Ninety days to a result that you can walk into your CPO's office (or your CFO's office, or your CEO's office) and say: "Here's what AI did for us. Here's the specific value it created. And here's why we should do more of it."

The quick win serves two purposes.

First, it demonstrates value—real, measurable, attributable value that justifies continued and expanded investment.

Second, it creates advocates—procurement professionals who have personally experienced the capability and who become your most effective internal champions for broader adoption.

So where do you start? Based on my experience and the capabilities we've explored throughout Part Two, there are three implementation starting points that consistently produce quick wins.

Option A: Negotiation Preparation (Highest Impact, Lowest Barrier)

Start with a single significant negotiation that's coming up in the next sixty to ninety days. Apply the full Chapter Seven preparation framework—market intelligence, supplier analysis, contract review, creative options generation, war-gaming. Measure the outcome against comparable negotiations that were prepared traditionally. Report the results.

This starting point works because it requires no technology investment beyond a general-purpose AI subscription, no data integration, no platform implementation, and no organizational restructuring. It requires one procurement professional willing to try a new approach on one negotiation. And if the framework produces a measurably better outcome (which, in my experience, it consistently does) you have a compelling proof point that requires no interpretation or extrapolation.

I have seen this approach produce results that changed organizational trajectories. One category manager who applied AI-powered preparation to a $4 million logistics renewal achieved an outcome that was eleven percent better than the target her leadership had set—approximately $440,000 in incremental value. She presented the preparation approach and the results to her procurement leadership team. Within two months, three other category managers had adopted the same workflow. Within six months, the entire team was using AI-powered preparation as standard practice. No mandate required. The results did the convincing.

Option B: Spend Analysis (Broadest Impact, Moderate Effort)

If your organization has spend visibility challenges—and as we discussed in Chapter Four, most do—an AI-powered spend classification and analysis project is an excellent starting point. It produces broad organizational value (everyone benefits from better spend visibility), it demonstrates AI capability in a concrete and visual way (dashboards and insights that stakeholders can see and interact with), and it creates a data foundation that supports virtually every other AI application in procurement.

The effort here is moderate rather than minimal—you'll need to extract transaction data, work through the classification and normalization process, and validate the results. But with the workflow described in Chapter Four, a single procurement pro can produce a comprehensive spend analysis in a matter of weeks that would have taken a team months using traditional methods. The output—a clean, categorized, normalized view of organizational spending that reveals consolidation opportunities, compliance gaps, and sourcing priorities—is immediately actionable and immediately impressive to stakeholders and leadership.

Option C: Contract Intelligence (Highest Dollar Recovery Potential)

If your organization has a large contract portfolio that hasn't been systematically analyzed—and the story at the beginning of Chapter Eight resonated a little too much with your current reality—a contract intelligence initiative can produce significant quick wins. The discovery of missed auto-renewals, uncaptured benchmarking rights, misapplied escalation clauses, and unfavorable terms that should be renegotiated consistently produces recoverable value that funds the entire AI initiative many times over.

Start with your highest-value contracts—the top fifty or hundred by annual spend. Use AI to extract and analyze the key terms, benchmark them against market standards, and identify the highest-priority opportunities for value recovery or improvement. The $4 million benchmarking clause story from Chapter Eight isn't a one-time fluke—that kind of hidden value exists in most large contract portfolios, waiting to be discovered by someone with the tools to find it.

SCALING BEYOND THE QUICK WIN: THE PHASED ROADMAP

The quick win gets you started and builds organizational support. Scaling to enterprise-wide AI adoption requires a more structured approach—but one that maintains the quick-win mentality of delivering value incrementally rather than promising transformation someday.

Phase One: Foundation (Months 1-3)

Execute your quick win. Build the initial proof point. Identify the two or three procurement pros who are most enthusiastic about AI adoption and support them in developing their skills. Begin building the prompt libraries, workflow documentation, and training materials that will support broader adoption. Establish the basic governance framework—data security guidelines, validation requirements, and ethical use standards that we'll discuss in the next chapter.

Phase Two: Expansion (Months 4-9)

Expand AI adoption from the initial pioneers to the broader team. This is where the training investment matters most—not classroom training about AI concepts, but hands-on workshops where sourcing professionals apply AI tools to their actual work with guidance from colleagues who have already developed proficiency. Roll out the standardized workflows for negotiation preparation, market intelligence, spend analysis, and contract review. Begin measuring the outcome metrics described in Chapter Nine.

This is also the phase where you should be evaluating purpose-built procurement AI tools. The quick win was likely achieved with general-purpose AI tools—ChatGPT, Claude, or similar. As adoption scales, the question of whether to invest in specialized platforms (spend analytics tools, contract intelligence platforms, risk monitoring systems) becomes relevant. The evaluation should be grounded in the specific use cases where general-purpose tools have proven valuable but where specialized tools could deliver more—better accuracy, better integration with your data systems, better workflow automation.

Phase Three: Integration (Months 10-18)

This is where AI moves from a set of tools that the best buyers use alongside their existing workflows to an integrated capability embedded in how procurement operates. Spend analytics are running continuously rather than being generated periodically. Risk monitoring is providing ongoing alerts rather than being checked manually. Contract intelligence is tracking renewals and obligations automatically rather than being consulted on demand. The AI-augmented operating model described in Chapter Nine is fully operational.

This phase typically requires more significant technology investment (platform subscriptions, data integrations, possibly custom development) and more formal change management. But by this point, you have twelve to eighteen months of demonstrated value, a team of procurement professionals who are skilled and enthusiastic users, and a set of business cases grounded in actual results rather than projected benefits. The investment conversation is significantly easier when you can show what AI has already delivered and make a credible case for what deeper integration would add.

Phase Four: Optimization (Ongoing)

AI capabilities are evolving rapidly. The tools available today are meaningfully better than those available a year ago, and the tools available a year from now will be meaningfully better than today's. Phase four is the ongoing practice of monitoring the AI landscape for new capabilities, evaluating their relevance to your procurement practice, and continuously improving your workflows to incorporate the best available tools and approaches.

This isn't a phase that ends. It's a permanent posture of informed adaptation—staying current with the technology, experimenting with new capabilities, and continuously raising the bar for what AI-augmented procurement can deliver.

BUILDING THE BUSINESS CASE: SPEAKING THE LANGUAGE OF INVESTMENT

Let me give you a practical framework for building the business case for AI investment in procurement, because this is where many initiatives stall—not for lack of value, but for lack of a compelling financial narrative.

The business case for AI in procurement has three components.

Direct value creation.

This is the most straightforward: better negotiation outcomes, lower pricing through improved market intelligence, recovered value from contract intelligence, savings from spend analysis insights. Quantify these based on your quick win results and project them across the portfolio. If your quick win negotiation produced an eleven percent improvement on a $4 million contract, what would comparable improvements across your top twenty negotiations represent? If your contract intelligence analysis recovered $340,000 from a single misapplied escalation clause, what might a comprehensive portfolio review recover?

Efficiency gains.

The time savings from AI-augmented workflows translate into either headcount efficiency (the ability to manage the same or greater scope with the existing team) or capacity reallocation (the ability to redirect time from low-value analytical work to high-value strategic work). Quantify the time savings from workflows—the spend analysis that takes days instead of weeks, the negotiation preparation that takes hours instead of days—and translate them into financial terms.

Risk mitigation.

The risk monitoring and contract intelligence capabilities described in Chapters Seven and Nine produce value that's harder to quantify but often the most significant. The disruption that was avoided because early warning gave you time to prepare. The auto-renewal that was caught before the deadline passed. The supplier financial deterioration that was detected months before a service failure. Frame these in terms of the cost of the risks they mitigate —the cost of a supplier failure, the cost of a missed renewal, the cost of a supply chain disruption—and the probability reduction that AI monitoring provides.

When you combine these three components (direct value, efficiency, and risk mitigation) the business case is typically overwhelming. The challenge is usually not that the value isn't there, but that it's spread across many small improvements rather than concentrated in a single dramatic number. The art of the business case is aggregating those improvements into a narrative that demonstrates the cumulative impact.

THE DATA FOUNDATION: GETTING YOUR HOUSE IN ORDER

I've deliberately avoided making this chapter a technology implementation guide—there are enough of those, and they age quickly as the technology evolves. But there is one technical topic that deserves attention because it is the single most common barrier to successful AI implementation in procurement:

Data quality.

Every AI capability we've discussed in this book depends on data. Spend analysis depends on transaction data. Supplier intelligence depends on financial and operational data. Contract intelligence depends on contract document accessibility. Risk monitoring depends on multi-source data feeds. And the quality of the AI's output is directly, inescapably constrained by the quality of the data it's working with.

Most procurement organizations have significant data quality challenges. Spend data is fragmented across multiple systems with inconsistent coding. Supplier data is duplicated, outdated, and poorly maintained. Contract documents are scattered across shared drives, email attachments, and individual procurement teams' local files. Performance data is collected inconsistently if at all.

You don't need perfect data to start using AI—as we discussed in Chapter Four, AI can actually help improve data quality through automated classification and normalization. But you do need accessible data. If your spend transactions are locked in an ERP system that nobody on the procurement team can query directly, AI can't help you analyze them. If your contracts are in a physical filing cabinet in a storage room, AI can't help you extract their terms. If your supplier information is scattered across a dozen spreadsheets maintained by different procurement professionals, AI can't help you build a unified supplier view.

The minimum data requirements for getting started are straightforward:

1. Procurement transaction data exportable in a structured format
2. Access to your contract documents in digital form
3. A list of your suppliers with basic information about each one.

If you have those three things, you can start. If you don't, getting them should be your first priority—not because AI requires them, but because effective procurement requires them, with or without AI.

COMMON IMPLEMENTATION FAILURES AND HOW TO AVOID THEM

Let me name the failure modes I've seen most frequently, because avoiding them is at least as important as following best practices.

The boil-the-ocean failure. The organization tries to implement everything at once—spend analytics, contract intelligence, risk monitoring, negotiation support, all in parallel, all enterprise-wide, all integrated. The implementation becomes so complex that it takes years, costs millions, and delivers nothing usable until everything is "ready." By that time, the executive sponsors have changed, the budget has been cut, and the initiative is quietly shelved.

Avoid it: Start with one use case, one team, one quick win. Scale from proven value, not from projected value.

The technology-first failure. The organization selects and implements a platform before clearly defining the use cases it needs to support and the workflows it needs to enable. The platform is powerful and feature-rich, but it doesn't fit how the procurement team actually works. Procurement pros are forced to adapt their workflows to the platform's design rather than the platform being configured to support their workflows. Adoption stalls because the tool feels like an imposition rather than an enablement.

Avoid it: Start with the workflow, not the technology. Define what you're trying to accomplish before you select tools to accomplish it.

The ivory tower failure. The AI initiative is led by a technology team or an innovation group that doesn't include sourcing professionals. The tools and workflows they develop are technically impressive but practically disconnected from how procurement actually works. The best buyers feel that the AI initiative was designed for them without consulting them, and they resist accordingly.

Avoid it: Include procurement professionals in every stage of the implementation. The people who do the work should design how AI augments the work.

The training gap failure. The organization provides tools but not training. Procurement teams are given AI subscriptions and told to "explore." Most of them don't. The ones who do have inconsistent experiences because they haven't been taught effective prompting techniques. The organization concludes that AI isn't useful for procurement.

Avoid it: Invest in practical, hands-on training that shows procurement professionals exactly how to use AI for the specific tasks they perform. Budget as much for training as for technology.

The governance vacuum failure. The organization adopts AI tools without establishing guidelines for data security, output validation, or ethical use. A sourcing professional uploads confidential supplier pricing data to a public AI tool. Another procurement professional cites a hallucinated benchmark in a negotiation. A third uses AI to generate a supplier evaluation that reflects the model's biases rather than the supplier's actual capabilities. Each of these incidents erodes trust in AI and creates organizational backlash.

Avoid it: Establish clear, simple governance guidelines before broad adoption. We'll cover this in detail in the next chapter.

The pattern across these failure modes is consistent: organizations that scale ambition faster than they scale capability tend to flame out. The organizations that navigate implementation successfully share the opposite trait—they are pragmatic rather than perfectionist.

A REAL IMPLEMENTATION STORY: FROM SKEPTICISM TO STANDARD PRACTICE

Let me walk you through an implementation I was closely involved with (not a composite, but a real story with identifying details changed) because the theory of implementation is less useful than watching one unfold with all its messiness, setbacks, and eventual success.

The organization was a mid-cap pharmaceutical company. Their procurement function was mature by traditional standards—good sourcing methodologies, competent procurement pros, a reasonable technology stack including an S2P platform that was five years old and showing its age. The CPO was forward-thinking and had secured budget for an "AI transformation initiative." The CIO was supportive but cautious about data security. The procurement team of thirty-five people ranged from enthusiastic early adopters to openly skeptical veterans who had survived two previous "transformation" initiatives that had produced a lot of PowerPoint and very little change.

Month One: The false start.

The CPO's initial plan was ambitious—implement an AI-powered spend analytics platform, a contract intelligence tool, and a risk monitoring system simultaneously, with a twelve-month rollout and a consulting firm managing the implementation.

I talked them out of it.

Not because the vision was wrong, but because the

approach was almost guaranteed to produce the boil-the-ocean failure I described earlier. Too much scope. Too long a timeline before anyone sees value. Too much dependence on consultants who would leave before the organization had internalized the capability.

Instead, we agreed on a single quick win: one category manager, one upcoming negotiation, the Chapter Seven preparation framework, and a general-purpose AI tool that required zero IT implementation.

Month Two: The proof point.

The category manager—a fifteen-year veteran named Maria who was the kind of experienced, skeptical procurement professional whose endorsement would matter most —prepared for her logistics contract renewal using the AI-powered framework. I coached her through the process personally. The supplier dossier took her about three hours to build. She later told me it would have taken her team two weeks to produce comparable analysis manually.

She discovered that the supplier had recently lost two accounts in her region, that their earnings call revealed margin pressure in the exact division that served her company, and that their proposed eight percent increase was roughly double what the underlying cost drivers justified.

The negotiation produced a result that was four hundred thousand dollars better than the target her leadership had set for the renewal.

Maria, who had walked into the process skeptical that "some chatbot" was going to help her negotiate better than thirty years of experience, became the initiative's most vocal and credible advocate. When she presented her preparation approach and results to the broader procurement team, you could see the skepticism in the room shift to curiosity. Not because a technology vendor had shown them a demo. Because Maria (someone they respected, someone who had been as skeptical as they were) had used the tools on a real negotiation and gotten a measurably better result.

Months Three through Six: The expansion.

Building on Maria's proof point, we expanded in concentric circles. Five additional category managers adopted the negotiation preparation framework for their upcoming renewals. We ran a two-day training workshop using the Tier Two curriculum I described earlier. We started a monthly AI practice group where sourcing professionals shared what they were learning—what worked, what didn't, what prompts produced the best results, where the AI had gotten things wrong and how they'd caught it.

The results accumulated. One category manager discovered a benchmarking clause in a technology contract that nobody knew existed—nearly identical to the Chapter Eight story. Another used AI-powered market intelligence to identify an alternative supplier in a category that had been treated as sole-source for seven years, creating competitive leverage that produced a twelve percent price reduction. A third used spend analysis to discover that the organization was spending $1.8 million annually with a supplier that had been "terminated" two years earlier—the termination notice had been sent but the AP team had never stopped processing their invoices.

Each of these stories became fuel for broader adoption. Not mandated adoption—organic adoption, driven by the best buyers who saw their colleagues getting better results and wanted the same advantage.

Months Six through Twelve: The integration.

With proven value and broad procurement pro buy-in, the organization was ready for more significant technology investment. They evaluated purpose-built spend analytics and contract intelligence platforms with procurement professionals, not just IT driving the evaluation. They selected tools that integrated with their existing S2P platform. They configured the tools around the workflows the team had already developed using general-purpose AI, rather than forcing the team to adopt the vendor's prescribed workflows. And they established the data security guidelines, validation requirements, and ethical use standards that formalized the practices the early adopters had been developing informally.

By month twelve, the procurement function looked meaningfully different. Not because they had implemented a new technology platform—though they had. Because the procurement teams had changed how they worked. Category managers were spending eighty percent of their time on strategy and stakeholder engagement instead of data gathering. Negotiations were consistently more thoroughly prepared. Contract renewals were being flagged six months in advance instead of discovered six weeks before expiration. Spend visibility had expanded from about sixty percent of addressable spend to over ninety percent. And the CPO was having strategic conversations with the executive team that would have been impossible without the intelligence capabilities the team had built.

The total cost of the implementation—including technology subscriptions, training, and the CPO's time—**was less than the value generated by the first three negotiation wins**. The ROI was achieved before the formal technology implementation was even complete.

That's what good implementation looks like. Not a grand transformation launched from a boardroom. A pragmatic, value-oriented evolution that starts with one sourcing professional, one negotiation, and one proof point—and scales from there based on demonstrated results.

THE VENDOR EVALUATION FRAMEWORK: CHOOSING PURPOSE-BUILT TOOLS

When you're ready to move beyond general-purpose AI tools to purpose-built procurement AI platforms (and you should get there eventually, because purpose-built tools offer capabilities that general-purpose tools can't match) here's the evaluation framework I recommend.

Does it solve a real, painful problem?

Start with the pain, not the technology.

If your biggest challenge is spend visibility, evaluate spend analytics platforms.

If it's contract management, evaluate contract intelligence tools.

If it's supplier risk, evaluate risk monitoring platforms.

Don't buy a comprehensive suite because a vendor convinced you that you need comprehensive AI capabilities. Buy the tool that solves the problem that's actually costing you money.

Does it work with your data?

The most sophisticated AI platform in the world is useless if it can't access and process your data. Evaluate integration capabilities critically—can it connect to your ERP, your S2P platform, your contract repository? How much data preparation is required before the tool can produce useful output? How long does the initial implementation take? The vendor will tell you "six to eight weeks." The reality is usually longer. Talk to reference customers and ask specifically about implementation timeline and data integration effort.

Is the AI transparent?

Can you understand why the tool produced a particular result? If a spend analytics platform classifies a transaction into a specific category, can you see why? If a risk monitoring tool flags a supplier, can you understand what signals triggered the alert?

Black box AI, tools that produce results without explainable reasoning, should be treated with extreme caution in procurement, where decisions need to be defensible and auditable.

Does the vendor understand procurement?

This might seem obvious, but a disturbing number of "procurement AI" products are built by technology companies that have never worked in procurement. The difference shows—in the workflows that don't match how procurement actually operates, in the terminology that doesn't match how procurement professionals actually talk, in the features that solve problems that look important from the outside but aren't the ones that procurement pros actually struggle with. Ask the vendor about their team's procurement experience. Talk to their implementation consultants and assess whether they understand your work or are just configuring software. Better yet, have your sourcing professionals evaluate the tool—not your IT team.

What does the pricing model look like over three years?

Some procurement AI platforms have aggressive initial pricing that escalates significantly at renewal—which is ironic, given that helping you avoid exactly this kind of commercial dynamic is supposed to be what they do. Evaluate the total cost of ownership over a three-year period, including subscription fees, implementation costs, training, data integration, and ongoing support. And negotiate the contract the way you'd negotiate any strategic technology agreement—with the preparation framework from Chapter Seven.

THE ORGANIZATIONAL POLITICS OF AI ADOPTION

I would be dishonest if I wrote an implementation chapter without addressing organizational politics—because politics kill more AI initiatives than technology failures do.

The most common political challenge is the IT-procurement turf war. IT often sees AI as their domain—a technology initiative that should be led, governed, and controlled by the technology function. Procurement sees AI as a business capability that should be directed by the people who understand the business problems it needs to solve.

Both perspectives have merit.

The resolution is usually a partnership model where IT provides the technology infrastructure, security governance, and integration support, while procurement provides the use case definition, workflow design, and procurement professional training. Neither function can do this alone. The organizations where IT and procurement collaborate effectively on AI adoption outperform the ones where the two functions compete for ownership.

The second common political challenge is budget. AI tools cost money—subscriptions, implementation, training, ongoing support. That money has to come from somewhere, and in most organizations, the procurement budget is already tight. The business case framework I described earlier in this chapter is essential for securing investment.

But the business case alone isn't sufficient—you also need an executive sponsor who will champion the investment through the budget process and defend it when competing priorities emerge.

The third political challenge is the "not invented here" problem. Some procurement organizations resist AI tools that weren't built internally—they want to develop their own capabilities using internal data science teams. While there are scenarios where custom development makes sense, in most cases the build-versus-buy math strongly favors buying purpose-built tools and configuring them to your needs. The time and cost of developing procurement AI capabilities from scratch is significant, and the purpose-built vendors have a multi-year head start in model training, workflow optimization, and edge case handling.

Navigate these politics proactively. Identify the potential blockers before you start. Build the alliances you need. Get your executive sponsor committed and briefed. And frame the initiative in terms that resonate with each stakeholder's priorities—IT cares about security and integration, finance cares about ROI and cost control, executive leadership cares about competitive advantage and risk mitigation. Same initiative, different pitch for each audience.

Implementation isn't a technology problem. It's a discipline problem. The discipline of starting small when every instinct says start big. The discipline of measuring outcomes when measuring activity is easier. The discipline of including the best buyers who will use the tools in the decisions about which tools to use. The discipline of investing in people as much as technology. The discipline of letting demonstrated value drive scale rather than letting projected value justify scope.

The procurement leaders who internalize this discipline will deliver results that redefine what their organizations expect from procurement. The leaders who default to the comprehensive transformation playbook will produce the spectacular flameouts I described at the beginning of this chapter.

Someday is not a strategy. Start now. Start small. Start with value.

Now let's talk about something the rush to adopt AI sometimes overlooks but that matters enormously for the long-term credibility and sustainability of AI in procurement: the ethical framework that should govern how we use these powerful tools.

CHAPTER 11
ETHICS, BIAS, AND THE RESPONSIBLE USE OF AI IN PROCUREMENT

I need to tell you about a supplier evaluation that almost went very badly wrong.

A procurement team I was advising had used an AI tool to score and rank suppliers for a competitive sourcing event in a professional services category. The AI analyzed the proposals, the supplier qualifications, the financial data, and the references, and produced a ranked list of six finalists with detailed scoring across twelve evaluation criteria. The analysis was impressive—thorough, consistent, well-documented. The team was ready to present it to the steering committee as the basis for their recommendation.

I asked to see the underlying scoring before the presentation. Something had been nagging at me about the results, though I couldn't immediately articulate what. When I looked at the scores by criterion, I found the problem.

Two of the six finalists were small, minority-owned

businesses. They had submitted strong proposals—competitive pricing, relevant experience, positive references. But the AI's scoring had systematically undervalued them on a criterion called "organizational stability"—a factor that assessed the supplier's size, tenure, financial resources, and market position. The AI had been trained on historical evaluation data from the organization, and that historical data reflected a pattern: for years, the organization had awarded contracts overwhelmingly to large, established firms, and the evaluation scores in the training data reflected that preference. The AI had learned the pattern and was reproducing it—scoring smaller firms lower on "stability" not because they were actually less stable, but because the historical data said that firms of their size and profile typically scored lower.

The two minority-owned firms weren't unstable. One had been in business for fifteen years with zero contract defaults. The other had a stronger balance sheet, relative to their revenue, than two of the four larger finalists. But the AI's score didn't reflect that reality. It reflected a historical bias that the organization had been unconsciously perpetuating for years—and that the AI was now systematically encoding and amplifying.

We caught it before the presentation. We adjusted the scoring methodology, re-evaluated the suppliers on objective financial metrics rather than pattern-matched "stability" scores, and the final ranking changed materially—one of the minority-owned firms moved from fifth to second. They ultimately won the award based on the revised, bias-corrected evaluation.

But here's what keeps me up at night about that story:

What if nobody had checked? What if the AI's beautifully formatted, impressively detailed, apparently objective scoring had been presented to the steering committee without scrutiny? The biased result would have looked like rigorous analysis. The discrimination would have been invisible—laundered through an algorithm that nobody thought to question because the output looked so professional.

This is why ethics in AI-powered procurement matters. Not as an academic exercise. Not as a compliance checkbox. As a professional imperative that has real consequences for real people and real organizations.

Ethics in AI procurement gets framed as a set of guardrails that slow down the process.

That framing is wrong.

Ethics is a strategic capability that protects the organization from a specific failure mode the AI age has made dangerously easy to encounter: the bias-laundered-through-algorithm decision that would never have passed scrutiny in a human review but that gets approved because the output looks rigorous, objective, and data-driven. The five-principle framework that follows in this chapter (human accountability, bias auditing, transparency, verification before action, and continuous learning) is not a set of constraints on AI use. It's the operating discipline that makes AI use defensible. The procurement leaders who treat it as such will protect their organizations and their reputations. The leaders who treat it as compliance will eventually have to explain a decision they cannot defend.

THE BIAS PROBLEM: DEEPER THAN YOU THINK

The supplier evaluation example I just described illustrates one form of AI bias—historical bias, where the model reproduces patterns from training data that reflect past discrimination or suboptimal decision-making. But the bias problem in procurement AI is broader and more multifaceted than any single example can capture.

Training data bias.

AI models learn from data. If that data reflects historical biases (in supplier selection, in pricing, in performance evaluation, in geographic preference) the model will reproduce those biases. An AI trained on a decade of sourcing decisions from an organization that systematically favored domestic suppliers will learn to downrank international suppliers. An AI trained on performance data that was collected inconsistently across suppliers, with more scrutiny applied to smaller or newer suppliers, will learn to rate those suppliers as higher risk.

Proxy variable bias.

Even when you remove obviously discriminatory variables from an AI model, the model can find proxy variables that correlate with the removed factor and produce similar discriminatory outcomes. Remove supplier size as an explicit evaluation criterion, and the model might weight "years in business" or "number of office locations"—variables that correlate with size and produce similar results. This is subtle, difficult to detect, and particularly insidious because it creates the appearance of objectivity while perpetuating the same underlying biases.

Confirmation bias amplification.

AI tools that assist with supplier research and market intelligence can amplify human confirmation biases by providing information that supports existing assumptions. If you ask an AI to "evaluate whether Supplier A is a good fit for our needs," the AI—which is designed to be helpful—may weight its analysis toward confirming that Supplier A is indeed a good fit, particularly if the framing of your question implies that expectation. A more neutral prompt—"evaluate the strengths and weaknesses of Supplier A relative to our requirements, and identify the three most significant risks of selecting them"—produces a more balanced analysis.

Exclusion bias.

AI systems work with available data. Suppliers who have less digital presence (fewer published case studies, less analyst coverage, smaller online footprint) are systematically disadvantaged in AI-powered research because there's simply less information for the AI to find and synthesize. This can disproportionately affect smaller suppliers, suppliers in developing markets, and diverse-owned businesses that may not have invested as heavily in the marketing and public relations activities that generate the digital footprint AI depends on.

THE TRANSPARENCY IMPERATIVE

When an experienced procurement professional makes a sourcing recommendation, the decision-making process is, at least in principle, transparent. You can ask them why they recommended Supplier A over Supplier B, and they can explain their reasoning: the market analysis they conducted, the evaluation criteria they applied, the trade-offs they considered, and the judgment calls they made. You might disagree with their reasoning, but you can understand it and challenge it.

When an AI system produces a recommendation, the transparency can be significantly lower. Machine learning models, particularly complex ones, can be true black boxes—producing outputs that even their creators can't fully explain. "The model scored Supplier A at 87 and Supplier B at 82" tells you the result but not the reasoning. And in procurement (where sourcing decisions involve significant financial commitments, affect supplier livelihoods, and must withstand internal and sometimes external scrutiny) the inability to explain why a recommendation was made is a serious problem.

The practical implication for procurement professionals is this: AI-generated recommendations should always be explainable. You should be able to articulate, in plain language, why the AI produced the result it produced and why you agree (or disagree) with that result. If you can't explain the AI's reasoning, if the model is truly a black box, you should not present its output as a recommendation. You should treat it as one input among many, and you should ensure that your recommendation is grounded in reasoning that you can explain and defend.

The principle that every AI-assisted procurement decision should be explainable by the human professional who is responsible for it is the cornerstone of responsible AI use in procurement.

It ensures accountability. It preserves the professional judgment that makes procurement a profession rather than a processing function. And it protects your organization from the legal, reputational, and ethical risks of making consequential decisions based on algorithmic outputs that nobody understands.

DATA PRIVACY AND CONFIDENTIALITY: THE RULES OF THE ROAD

We touched on data security in Chapter Three, but the ethical dimensions deserve deeper attention here. Procurement professionals handle a significant amount of sensitive information: supplier pricing data, internal cost targets, contract terms, strategic plans, financial information, and sometimes personally identifiable information about supplier employees.

When this information is shared with AI tools (uploaded for analysis, pasted into prompts, or processed through cloud-based platforms) several ethical and legal obligations come into play.

Confidentiality obligations.

Many supplier relationships are governed by non-disclosure agreements that restrict how information can be shared and used. If a supplier has shared their cost structure with you under NDA, uploading that cost structure to a public AI tool may violate the NDA—even if you're only using it for internal analysis. Know what's covered by confidentiality agreements and ensure your AI usage practices are consistent with those obligations.

Data protection regulations.

If any of the data you're working with includes personal information (supplier contact details, employee names, biographical data from LinkedIn research) it may be subject to data protection regulations like GDPR in Europe or CCPA in California. These regulations impose restrictions on how personal data can be processed, stored, and shared. AI tools that process data through cloud infrastructure may involve cross-border data transfers that trigger additional regulatory requirements. Understand the regulatory framework that applies to your data and ensure your AI usage is compliant.

Proprietary data training risks.

Some AI platforms use the data submitted by users to train and improve their models. This means that information you upload today could influence the model's outputs to other users tomorrow—potentially exposing proprietary information in ways you didn't intend. Most enterprise-grade AI offerings now provide options where your data is not used for model training, but you need to verify this rather than assume it. Read the terms of service. Ask the vendor directly. And if you can't get a clear, satisfactory answer about data handling, don't upload sensitive information.

The ethical framework here is straightforward: treat AI tools with the same data security discipline you would apply to any external service provider who handles your organization's sensitive information. Because that's exactly what they are.

BUILDING AN AI ETHICS FRAMEWORK FOR PROCUREMENT

Rather than treating AI ethics as a set of abstract principles, let me propose a practical framework of five principles that procurement organizations can adopt and adapt. Each one is operational rather than aspirational—designed to be applied in the daily work of an AI-augmented procurement team.

Principle One: Human accountability for every decision.

No procurement decision should be attributed to "the AI recommended it." Every sourcing recommendation, every supplier evaluation, every negotiation strategy should have a named human professional who is responsible for the decision and who can explain the reasoning behind it.

AI informs decisions. Humans make them. And humans are accountable for them.

Principle Two: Bias auditing as a standard practice.

Any AI-powered evaluation or scoring process (supplier evaluations, risk assessments, spend classifications) should be periodically audited for bias. This means examining the results for patterns that might indicate systematic disadvantage to particular categories of suppliers—by size, geography, ownership characteristics, or other factors. The bias audit doesn't need to be a massive undertaking. A quarterly review of AI-generated evaluations, looking specifically for patterns of systematic disadvantage, is sufficient to catch most bias issues before they become entrenched.

Principle Three: Transparency with stakeholders.

When AI tools are used to support procurement decisions, that fact should be disclosed to relevant stakeholders—including, in appropriate contexts, suppliers. This doesn't mean providing a detailed explanation of the AI methodology. It means being honest that AI-powered analysis was part of the decision-making process, and being willing to explain the reasoning behind the decision in terms that don't rely on "the algorithm said so."

Principle Four: Verification before action.

Every AI-generated output that informs a consequential decision should be verified by a human professional with relevant domain expertise. The specific verification requirements should scale with the stakes: a quick sanity check for routine analysis, rigorous source verification for data that will be cited in a high-stakes negotiation, and comprehensive validation for any AI output that informs a legally significant decision.

Principle Five: Continuous learning and adaptation.

AI ethics is not a static framework—it's an evolving practice. The ethical challenges that AI creates in procurement are still being discovered and understood. Commit to staying current with emerging best practices, learning from incidents (both your own and others'), and updating your ethical framework as the technology and the regulatory environment evolve.

THE SUPPLIER DIVERSITY DIMENSION

I want to give specific attention to supplier diversity because it sits at the intersection of AI bias, organizational values, and real business impact in ways that deserve careful consideration.

Most large organizations have supplier diversity programs—commitments to directing a portion of procurement spend to businesses owned by minorities, women, veterans, people with disabilities, and other underrepresented groups. These programs exist because supplier diversity is both a social good and a business advantage:

diverse supplier bases are more innovative, more resilient, and more aligned with the diverse customer bases that most organizations serve.

AI creates both risks and opportunities for supplier diversity.

We've discussed the risks above: bias in evaluation, exclusion from AI-powered research, and systematically lower scoring on criteria that correlate with firm size.

But the opportunities are equally significant.

AI can enhance supplier diversity by making it easier to discover qualified diverse suppliers that traditional research methods might miss. AI-powered market mapping can identify diverse-owned businesses in categories where the procurement team wasn't aware they existed. AI can analyze spend data to identify categories where diverse supplier participation is low and where targeted outreach could increase it. AI can benchmark the organization's diversity performance against industry standards and track progress over time.

The key is intentionality. If you simply deploy AI tools without explicitly incorporating diversity objectives, the default outcome will likely be reduced diversity, because the AI's training data reflects a world where diverse suppliers are underrepresented. The model will reproduce that underrepresentation unless deliberately directed otherwise.

If you intentionally incorporate diversity criteria into your AI-powered processes by including diverse suppliers in market mapping, adjusting evaluation criteria to avoid size-correlated bias, and using AI to identify diversity opportunities rather than just efficiency opportunities—the technology can accelerate your diversity goals rather than undermining them.

THE BIGGER PICTURE: AI ETHICS AS PROFESSIONAL ETHICS

The ethical framework for AI in procurement is not a separate, new set of rules. It's an extension of the professional ethics that have always governed great procurement practice.

The procurement professional's ethical obligations haven't changed because AI arrived. The obligation to be honest with suppliers hasn't changed. The obligation to run fair, objective sourcing processes hasn't changed. The obligation to protect confidential information hasn't changed. The obligation to make decisions based on merit rather than bias hasn't changed.

What has changed is the set of tools through which those obligations are fulfilled or violated. AI creates new mechanisms for bias, new risks of data exposure, new possibilities for unfair advantage, and new challenges for transparency. But the ethical principles that govern the response to those challenges are the same principles that have always governed professional procurement practice: honesty, fairness, transparency, accountability, and the disciplined application of professional judgment.

The procurement professionals who approach AI ethics with the same seriousness they bring to negotiation ethics, supplier relationship ethics, and sourcing process ethics will navigate the challenges effectively. The procurement teams who treat AI ethics as somebody else's problem will eventually have to defend a decision they cannot defend—because the practical scenarios where ethical AI use matters most are not abstract, and the next two sections describe them directly.

WHEN AI AND HUMAN JUDGMENT DISAGREE: A PRACTICAL FRAMEWORK

There's a scenario that every AI-adopting procurement professional will eventually face, and it deserves pointed attention: the moment when the AI's analysis points in one direction and your professional judgment points in another.

Maybe the AI's supplier risk model flags a supplier as high risk based on financial signals, but your relationship with the supplier gives you context that the model can't see. You know they just closed a major funding round that hasn't been publicly announced yet, and the financial stress indicators are temporary. Maybe the AI's spend analysis identifies a consolidation opportunity in a category, but you know from experience that the two suppliers it suggests consolidating serve different business units with incompatible requirements. Maybe the AI's negotiation preparation recommends an aggressive opening position based on market benchmarks, but your relationship intelligence tells you that this supplier is a critical partner who will respond to aggression with disengagement rather than concessions.

In each of these cases, what do you do?

The wrong answer is to defer to the AI because it's

"data-driven" and your judgment is "subjective." The whole point of Part Two was that AI provides analysis while humans provide judgment. When they conflict, the human judgment should prevail—but *only* if the judgment is informed rather than reflexive.

Here's my practical framework for handling AI-human disagreement:

Step one: understand why the AI reached its conclusion.

Before overriding the AI's analysis, make sure you understand what data and patterns drove the result. The AI might be seeing something you're not. That supplier risk flag might be based on signals you haven't considered, like supplier employee reviews trending negative, a pattern of late payments to their own suppliers, or a subtle decline in patent filing activity. Understanding the AI's reasoning might actually change your judgment rather than confirming it.

Step two: articulate why your judgment differs.

If, after understanding the AI's reasoning, you still disagree, articulate specifically why. "I have a feeling the supplier is fine" is not sufficient. "The supplier's CFO told me directly in our quarterly review that they secured a $50 million credit facility last month, which addresses the liquidity concern the model is flagging" is sufficient. The discipline of articulating your reasoning (to yourself, not just to others) prevents the lazy override of AI analysis based on gut feel or cognitive bias.

Step three: document the disagreement and your reasoning.

This serves two purposes. First, it creates accountability—if your judgment turns out to be wrong, the documentation helps you and the organization learn from it. Second, it provides data that can improve the AI model over time. If procurement professionals consistently override the AI's risk flags for a specific type of signal, that pattern suggests the model needs to be retrained or the signal needs to be weighted differently.

Step four: monitor the outcome.

After you've overridden the AI's recommendation, track what actually happens. Was the supplier fine, as you expected? Or did the risk materialize, as the AI predicted? Over time, this monitoring creates a feedback loop that calibrates both the AI model and your own judgment—showing you where the AI tends to be right and where your experience-based insight adds real value.

The goal is not AI supremacy or human supremacy. It's a calibrated partnership where each party's strengths compensate for the other's weaknesses. The AI catches patterns in large data sets that humans miss. Humans provide context, relationship intelligence, and organizational judgment that AI can't access. The best outcomes come from neither blind deference to the algorithm nor stubborn insistence on gut feel, but from a disciplined dialogue between the two.

THE REGULATORY LANDSCAPE: WHAT'S COMING AND WHY IT MATTERS

The regulatory environment for AI is evolving rapidly, and procurement professionals need to be at least directionally aware of what's coming, because AI regulations are going to affect how you can use AI tools, what governance your organization needs to have in place, and what obligations you may have to suppliers and other parties.

The most significant regulatory development is the European Union's AI Act, which entered into force in 2024 and has been implemented in phases since. The AI Act establishes a risk-based framework for AI governance, which categorizes AI applications by risk level and imposes progressively stricter requirements for higher-risk applications. Most procurement AI applications are likely to fall into the "limited risk" category, which requires transparency obligations (telling people when they're interacting with AI) but not the more burdensome requirements that apply to high-risk applications.

In case you were wondering, there *are* scenarios where procurement AI could be classified as higher risk. AI systems used for evaluating suppliers could potentially be considered analogous to the employment evaluation systems that the AI Act classifies as high risk—particularly if the AI's recommendations directly affect supplier livelihoods or if the evaluation incorporates personal data about supplier employees. AI systems used for automated contract negotiation or autonomous purchasing decisions could raise similar questions. The regulatory interpretation continues to evolve, and procurement leaders should be monitoring developments in consultation with their legal teams.

Beyond the EU, other jurisdictions are developing their own AI regulations. The United States has taken a more sector-specific approach, with various federal agencies issuing guidance on AI use in their regulated domains. State-level regulations, particularly California's, are adding additional requirements. China has implemented its own AI governance framework. And international organizations are developing AI ethics principles that, while not legally binding, influence corporate governance practices globally.

The practical implication for procurement is this: build your AI governance framework now, before regulation forces you to. The organizations that have proactively established transparency practices, bias auditing, data governance, and human oversight for their AI-powered procurement processes will be well-positioned when regulations arrive. The organizations that wait until compliance is mandatory will be scrambling to retrofit governance onto AI practices that have been operating without it.

REAL-WORLD ETHICAL DILEMMAS: NO EASY ANSWERS

Let me present a few ethical scenarios that don't have clean answers—because part of ethical maturity is learning to navigate ambiguity rather than just following rules.

Scenario One:
The efficiency-diversity tension.

Your AI-powered spend analysis identifies a consolidation opportunity in a professional services category. Consolidating from eight suppliers to three would save approximately $2 million annually. But two of the five suppliers being eliminated are diverse-owned businesses that your organization has been deliberately supporting as part of its supplier diversity program. The diverse suppliers are slightly more expensive and slightly less efficient—which is exactly why the AI's optimization model recommends eliminating them. What do you do?

There's no clean answer. The savings are real and meaningful. The diversity commitment is also real and meaningful. The resolution requires a judgment call that weighs financial value against organizational values—exactly the kind of judgment that AI **cannot** and **should not** make.

Perhaps the answer is a modified consolidation that retains one of the diverse suppliers and provides development support to help them become more competitive. Perhaps it's accepting a slightly lower savings target in exchange for maintaining diversity commitments. Perhaps it's finding other categories where diverse supplier participation can be increased to offset the loss. Whatever the resolution, it should be a deliberate, transparent human decision—not a default outcome of algorithmic optimization.

Scenario Two:
The competitive intelligence boundary.

Your AI-powered supplier research has identified that a key supplier's employee reviews on Glassdoor have been declining sharply, with specific complaints about leadership instability and missed pay dates. This is valuable intelligence for your risk assessment. It suggests the supplier may be experiencing financial or organizational distress. But some of those reviews may include information that the employees expected to be shared anonymously with potential job seekers, not synthesized by an AI tool and used in a commercial negotiation. Is it ethical to use this intelligence?

The information is publicly available—anyone with a web browser can read Glassdoor reviews. But the scale and synthesis capability of AI transforms the nature of the intelligence. Reading a few reviews casually is different from systematically analyzing hundreds of reviews to build a predictive model of organizational health. The information itself may be fair game, but the use of it raises questions about proportionality and intent that deserve thoughtful consideration.

Scenario Three:
The automation displacement question.

Your AI implementation is going well—procurement

pros are more productive, outcomes are improving, the function is delivering more value. But the efficiency gains are real enough that your organization could manage its current procurement scope with twenty-five percent fewer people. Your CFO has noticed. They're asking whether the headcount reduction should be captured as a cost saving. What's your ethical obligation to the team members whose roles might be eliminated?

This is not a hypothetical. It's a conversation happening in procurement organizations right now.

The ethical answer, in my view, is that leaders have an obligation to be transparent with their teams about how AI will affect roles, to invest in reskilling and redeployment before resorting to layoffs, and to advocate for using efficiency gains to expand procurement's scope and impact rather than simply reducing headcount. The procurement function that uses AI to do the same work with fewer people has missed the point. The one that uses AI to do significantly more valuable work with the same people has understood it.

None of these scenarios has a simple answer. That's the point. Ethical practice in the AI era requires the same qualities it has always required: real commitment to doing the right thing, willingness to engage with ambiguity rather than avoiding it, and the professional judgment to navigate complex situations where multiple legitimate values are in tension.

The five principles I outlined earlier in this chapter (human accountability, bias auditing, transparency with stakeholders, verification before action, and continuous learning and adaptation) are the operating discipline that makes that judgment defensible. Internalize the principles. Apply them in the dilemmas. Bring them to the conversations where the easy thing and the right thing point in different directions. The framework is not a set of rules to follow when the answer is obvious. It's a structure that holds when the answer is not.

Ethics in procurement has always been about reputation—the reputation that, as we said in Chapter Two, is not soft but is leverage. AI doesn't change that. It just raises the stakes.

Now let's close this book the way we opened it—with a look at where this profession is headed, and what the sourcing professionals who are ready to lead it will need to do to get there.

CHAPTER 12
THE FUTURE OF PROCUREMENT

WHAT COMES NEXT AND WHO WILL LEAD IT

The year is 2030. Maybe 2032. Not so far away that it feels like science fiction, but far enough that the changes between now and then will be significant enough to reshape the procurement profession in ways that are difficult to predict with precision but possible to anticipate in broad strokes.

I want to use this final chapter not to make predictions —predictions about technology are almost always wrong, but to paint a picture of where the trajectory we've been discussing throughout this book leads. Because the transformation we've explored is not a one-time event. It's a compounding process. Each capability builds on the one before it. Each innovation enables the next. And the pace of change is accelerating, not stabilizing.

The procurement profession that emerges from this transformation will be recognizable to the best buyers of today—the fundamentals we discussed in Part One will endure. But the way those fundamentals are practiced, the tools that amplify them, and the expectations that organizations have for what procurement can deliver—those will be different in ways that are both exciting and challenging for the professionals who will navigate them.

Let me share what I think that future looks like.

A note for senior leaders before we go further:

This is the chapter that bookends Chapter One's argument that "procurement matters." The case has been made across the twelve chapters that followed: the fundamentals are durable, the AI capability is transformed, and the profession is in the middle of the most significant capability shift in its history.

The strategic question this chapter addresses is not whether the future I'm about to describe will arrive. It will. The question is whether your organization will be in the position to capture the value when it does. The procurement leaders who treat this chapter's projections as actionable strategic intelligence, not speculative futurism, will be the ones whose functions are positioned to lead. The ones who treat it as interesting reading will be the ones whose functions are positioned to follow.

THE AUTONOMOUS PROCUREMENT LAYER

The most significant near-term change is the emergence of what I'll call the autonomous procurement layer—the set of procurement activities that will be handled entirely by AI systems with minimal or no human intervention.

This is already happening in narrow domains. The autonomous negotiation capabilities we discussed in Chapter Three—where AI agents execute tactical sourcing events and negotiations for routine, well-defined categories—are operational today at organizations using platforms like Procure AI. They're handling RFQs for commodity purchases, conducting automated negotiations for tail spend, and generating purchase orders based on pre-defined rules and parameters. The results are measurable: faster cycle times, consistent execution, and savings capture on transactions that previously didn't receive any negotiation attention because they were too small to justify a human procurement professional's time.

Over the next five to ten years, this autonomous layer will expand significantly. Routine purchase requisitions will be processed, sourced, and fulfilled without human involvement—the AI will identify the need, match it to a preferred supplier and contract, validate the pricing, generate the order, and route it for automated approval if it meets pre-defined criteria. Invoice matching and payment processing will be fully automated for standard transactions. Routine contract renewals (the straightforward, low-risk renewals where the relationship is stable, the terms are market-competitive, and the strategic stakes are low) will be handled by AI agents that negotiate within pre-approved parameters and escalate to humans only when the negotiation falls outside those boundaries.

The key phrase in the previous paragraph is "routine." The autonomous layer handles the work that is predictable, rule-based, and low-risk. The moment a situation becomes novel, complex, politically sensitive, or strategically significant, the moment judgment is required, it escalates to a human professional.

This creates a procurement operating model with two distinct layers. The autonomous layer handles volume—the thousands of transactions, the hundreds of routine negotiations, the ongoing monitoring and compliance activities that consume significant human bandwidth today. The human layer handles value—the strategic sourcing decisions, the complex negotiations, the supplier relationship development, the risk management, and the stakeholder engagement that require the judgment, creativity, and interpersonal skill that we've discussed throughout this book.

The procurement professionals who thrive in this model are not the ones who can process transactions faster than the AI—that's a race they'll lose. They're the ones who can do what the AI can't: navigate ambiguity, build trust, exercise political judgment, create innovative solutions, and lead strategic change. The fundamentals from Part One. Amplified by the AI capabilities from Part Two. Applied in the organizational context from Part Three.

THE PREDICTIVE PROCUREMENT FUNCTION

Beyond autonomous execution of routine tasks, AI is moving procurement toward a truly predictive model—one where the function can anticipate needs, risks, and opportunities before they materialize rather than responding to them after the fact.

Imagine a procurement function that can predict, with reasonable accuracy, which categories are going to experience significant price increases in the next quarter (based on commodity market trends, supplier financial signals, demand patterns, and macroeconomic indicators) and proactively adjust its sourcing strategies before the increases hit.

That can predict which suppliers are at elevated risk of disruption, based on the early warning signals we discussed in Chapter Six, and proactively qualify alternatives before the disruption occurs.

That can predict which contracts are likely to experience cost overruns based on early spending patterns and intervene while there's still time to course-correct. That can forecast organizational demand for key categories and optimize procurement strategies around predicted needs rather than historical patterns.

This predictive capability is emerging now. The tools and models are still maturing, and the accuracy of predictions varies significantly depending on the category, the data quality, and the complexity of the market dynamics.

But the direction is clear:

Procurement is moving from a reactive function that responds to business needs and market events to a proactive function that anticipates and shapes them.

The implications for procurement teams are profound. The predictive procurement professional is not a forecaster —the AI handles the forecasting. They are a strategic translator—someone who can take the AI's predictions and translate them into organizational action.

"The model is showing a seventy percent probability of a fifteen percent increase in logistics costs next quarter due to a combination of fuel price trends and shipping capacity constraints. Here's the sourcing strategy I recommend to mitigate the impact, here are the three suppliers I've prequalified as alternatives, and here's the commercial framework I propose for locking in current rates through extended commitments."

That's a very different conversation from the one procurement has traditionally had with the business—which was typically some variation of "costs went up and here's why." Moving from explanation to anticipation, from reaction to preparation, changes the value proposition of the procurement function in ways that will reshape its organizational positioning.

THE INTELLIGENCE NETWORK

One of the most intriguing longer-term possibilities is what I'll call the procurement intelligence network—the emergence of shared intelligence capabilities across organizations that amplify the individual capabilities we've discussed throughout this book.

The concept is straightforward, even if the execution is complex. Individual procurement organizations generate significant amounts of intelligence through their AI-powered activities: market pricing data, supplier performance assessments, contract terms benchmarks, risk monitoring signals. Most of this intelligence is siloed within the individual organization. But if it could be aggregated, anonymized, and shared, the way that community benchmarking data is already shared through platforms like Coupa's business spend management community, the collective intelligence would be significantly more powerful than any individual organization's data set.

Imagine being able to benchmark your supplier pricing against your own historical data and anonymized pricing data from hundreds of organizations buying from the same suppliers in the same categories.

Imagine being able to assess supplier risk not just based on publicly available signals, but based on the aggregated performance experience of dozens of customers who work with the same supplier.

Imagine being able to identify market trends not from analyst reports that are published quarterly, but from real-time transaction data flowing across hundreds of organizations simultaneously.

This kind of network intelligence already exists in limited forms—community-sourced benchmarking, shared risk monitoring, collaborative supplier assessment programs. AI amplifies its potential by making it possible to aggregate, normalize, and analyze the data at a scale and speed that manual approaches can't match.

The procurement organizations that participate in these intelligence networks, that both contribute to and benefit from shared data, will have information advantages that isolated organizations cannot replicate. And the procurement professionals who understand how to leverage network intelligence, how to combine community data with their organization's context and needs, will be the most informed participants in any market they operate in.

THE EVOLVING CPO: FROM COST CONTROLLER TO VALUE ARCHITECT

The Chief Procurement Officer role is going to evolve significantly as AI transforms the function. The CPO of the future is not primarily a cost controller—though cost management remains important. They are what I'd call a value architect: someone who designs and orchestrates the systems, relationships, and capabilities that create sustainable competitive advantage through procurement.

The value architect CPO operates at the intersection of technology, strategy, and organization. They understand AI well enough to direct its application strategically. They understand markets well enough to identify the opportunities that AI reveals. They understand organizations well enough to build the teams, processes, and partnerships that translate opportunities into outcomes. And they understand risk well enough to navigate the ethical, operational, and strategic risks that AI creates alongside the value it enables.

This role requires a broader skill set than the traditional CPO role. It requires technology literacy—not the ability to code or configure algorithms, but the ability to evaluate AI capabilities, direct their application, and govern their use. It requires strategic communication—the ability to translate procurement's increasingly sophisticated capabilities into language that the CEO, CFO, and board of directors understand and value. It requires organizational leadership—the ability to manage the human dynamics of a function that is being significantly transformed by technology. And it requires ethical grounding—the ability to navigate the complex ethical questions that AI creates with the kind of principled judgment that builds trust rather than eroding it.

The CPOs who develop this integrated skill set (who combine procurement expertise, technology literacy, strategic thinking, and ethical leadership) will be among the most valuable executives in their organizations. Because the procurement function they lead will be delivering value that no other function can replicate: the intelligence, the relationships, the risk management, and the cost optimization that come from managing an organization's external resource acquisition with the kind of sophistication and capability that AI makes possible.

THE ENDURING HUMAN ELEMENTS

Throughout this book, I've been making a consistent argument: AI amplifies procurement expertise rather than replacing it. I've said this so many times that you might suspect I'm repeating it for emphasis. I am.

Because the single most important message in this book is not about AI. It's about the human skills and judgment that AI makes more valuable, not less.

Let me name the human elements that will endure—that will matter as much in 2035 as they do in 2025, regardless of how sophisticated the AI becomes.

Relationships.

The trust, the credibility, the genuine partnership that great procurement professionals build with their suppliers and their stakeholders—Chapter One's central thesis—is not being automated. The supplier who picks up the phone at 11 PM when you need them is not responding to an algorithm. They're responding to a relationship. The stakeholder who trusts procurement enough to execute a coordinated negotiation play—Chapter Two's SaaS story—is not trusting a technology platform. They're trusting a person.

AI can generate the intelligence that makes relationships more productive.

It cannot build the relationships themselves. That remains profoundly, irreducibly human work.

Judgment.

The ability to take imperfect information and make a

good decision (to weigh competing factors, navigate ambiguity, account for organizational context, and choose a course of action that balances multiple objectives) is not something AI does. AI provides analysis. Humans provide judgment. And the decisions that matter most in procurement (the strategic choices about where to source, how to structure deals, when to take risks, and how to build long-term competitive advantage) require judgment that no algorithm can replicate.

Creativity.

The negotiation breakthrough that creates value neither party expected. The sourcing strategy that approaches a category from an angle nobody considered. The supplier development initiative that builds a capability the market didn't offer. These are creative acts—products of human imagination, insight, and willingness to see possibilities where others see constraints. AI can generate options. But the creative leap that turns a good option into a breakthrough outcome—that's human.

Ethics.

The principled judgment about what's right—not just what's profitable, not just what's legal, but what's right—is a human responsibility that becomes more important, not less, as AI provides more powerful tools. The ethical framework we discussed in Chapter Eleven—the commitment to fairness, transparency, accountability, and respect for the people on both sides of every procurement transaction—is a human value system that must govern the use of technology, not be governed by it.

Curiosity.

The genuine interest in understanding markets, suppliers, technologies, and people—the curiosity that Chapter Two identified as the defining trait of the best negotiators—is also the defining trait of the best AI adopters. The procurement pros who approach AI with the same curiosity they bring to everything else—who are always experimenting, always learning, always looking for the next way to create value—are the ones who will continuously push the frontier of what AI-augmented procurement can accomplish. Always be curious. Never stop learning. If I had to compress thirty years of procurement experience into a single piece of advice, that's the one I'd give. To you, to my students, and to a younger version of myself.

A DAY IN THE LIFE: THE PROCUREMENT PROFESSIONAL OF 2030

Let me paint a picture of what a typical day might look like for a senior procurement professional five years from now—not as science fiction, but as a plausible extrapolation of the trajectory we've been discussing.

Sarah is a VP of Procurement at a mid-cap public company. She arrives at work (or, more likely, opens her laptop at home, since the hybrid work model has long since become permanent) and starts with her AI-generated morning briefing. Overnight, her risk monitoring system flagged three items: a credit downgrade for a tier-two input supplier in Southeast Asia, a regulatory change in the EU that affects compliance requirements for two of her business lines, and a news report about a labor dispute at a logistics provider's distribution hub in the Midwest.

For the credit downgrade, the AI has already assessed the potential impact on her supply chain, the tier-two supplier provides a critical specialty input used by one of her tier-one suppliers, and has generated a preliminary risk assessment with three mitigation options, ranked by cost and implementation speed. Sarah reviews the options, adds context that the AI doesn't have (she knows the tier-one supplier has been qualifying an alternative source for that input and is about six months from completion), adjusts the recommended approach accordingly, and forwards the updated assessment to her category manager with instructions to contact the tier-one supplier for a status update on the alternative source.

Total time: twelve minutes. Five years earlier, this analysis would have taken two days—if anyone had detected the tier-two credit downgrade at all, which they probably wouldn't have.

For the EU regulatory change, the AI has already cross-referenced the new requirements against her current packaging supplier contracts and identified two that will need amendments. It has drafted the amendment language, benchmarked the likely cost impact, and scheduled a briefing with the regulatory compliance team. Sarah reviews the analysis, confirms it's directionally correct, and approves the briefing request.

Total time: eight minutes.

For the logistics labor dispute, the AI has assessed the probability of a strike based on historical patterns at this provider (they've had three labor actions in the past decade, with an average duration of eleven days) and has identified alternative routing options that would maintain ninety percent of current delivery capacity at a seven percent cost premium. Sarah decides this doesn't require immediate action but asks the AI to monitor the situation and alert her if the probability of a strike exceeds fifty percent.

Total time: five minutes.

It's 8:25 AM. Sarah has assessed and responded to three supply chain risks that would have consumed most of a day in the pre-AI era. She spends the rest of her morning in a strategic planning session with her CEO and CFO, presenting a supply chain resilience assessment that her AI platform generated overnight—a comprehensive model showing the organization's exposure to twenty-three categories of supply chain risk, with probability-weighted financial impact estimates and recommended mitigation investments. The conversation is substantive, strategic, and anchored in data that would have been impossible to compile without AI. The CFO approves a $4 million investment in supply chain diversification based on the analysis.

After lunch, Sarah coaches one of her category managers through preparation for a complex technology negotiation using the Chapter Seven framework. The AI-generated supplier dossier is already built. They spend an hour refining the strategy, war-gaming the supplier's likely responses, and developing creative deal structures. The category manager walks out of the session confident, well-prepared, and ready for a negotiation that will determine whether the organization pays $12 million or $14 million for a critical technology platform over the next three years.

Sarah's afternoon includes a video call with a strategic supplier's CEO—a quarterly alignment conversation that exists because of the kind of genuine partnership described in Chapter One. They discuss the supplier's product roadmap, Sarah's organization's evolving requirements, and a joint innovation initiative they've been developing. The AI generated a briefing for this conversation too—the supplier's recent financial performance, their competitive positioning, their recent patent filings that suggest where their R&D is focused. Sarah uses this intelligence not as a weapon but as a foundation for an informed conversation between partners.

Her day ends with a brief review of the AI's automated sourcing activities. Three tactical sourcing events were executed autonomously today—routine commodity purchases handled by AI agents within pre-approved parameters. One is flagged for her review because the winning bid was from a supplier that doesn't meet the organization's diversity requirements. She overrides the autonomous recommendation and directs the award to the second-ranked bidder, who is a certified diverse supplier at a three percent price premium that she judges is justified by the organization's diversity commitments. She makes a note to adjust the AI's parameters to incorporate diversity requirements in future autonomous sourcing events.

Total for the day: three supply chain risks assessed and addressed, a $4 million investment approved based on AI-generated analysis, a complex negotiation strategy developed, a strategic supplier relationship deepened, and three tactical sourcing events executed—all before dinner.

That's not science fiction. Every capability I described in Sarah's day exists today in some form. What changes between now and 2030 is the integration—the seamless connection between risk monitoring, market intelligence, contract analysis, autonomous execution, and strategic decision support that turns individual AI tools into a comprehensive procurement intelligence platform.

THE CAREER IMPLICATIONS: WHAT THIS MEANS FOR YOU

The transformation we've been discussing has concrete implications for procurement careers—for compensation, for career paths, and for the skills that will command a premium in the market.

Compensation is diverging.

The procurement professionals who have developed AI-augmented capabilities are commanding a significant premium in the market—because they're demonstrably more productive, more effective, and more strategically valuable than their peers who haven't. This premium will only grow as AI adoption becomes the expectation rather than the exception. The sourcing professionals who wait too long to develop AI skills will find themselves competing for a shrinking pool of roles that don't require them—roles that are, by definition, the least strategic and lowest-compensated in the function.

Career paths are broadening.

The traditional procurement career path, analyst to buyer to category manager to director to VP to CPO, is being supplemented by new paths that reflect the expanded capabilities of the AI-augmented function. Procurement data and analytics roles. AI implementation and governance roles. Strategic intelligence roles that combine market research, risk monitoring, and supplier intelligence into a dedicated function. And hybrid roles that bridge procurement and technology—the kind of role that someone who has built a procurement AI platform like ProcureOptima occupies, combining deep domain expertise with technology capability in a way that neither a pure procurement professional nor a pure technologist could replicate.

The CPO pipeline is changing.

The CPOs being hired today, and the ones who will be hired five years from now, increasingly need technology literacy alongside traditional procurement expertise. Boards and CEOs are looking for procurement leaders who can articulate how AI transforms the function, who can evaluate technology investments strategically, who can build AI-augmented teams, and who can leverage data and analytics to have the kind of strategic conversations that elevate procurement from a cost management function to a competitive advantage engine. The CPO candidate who can demonstrate AI-augmented procurement leadership has a meaningful advantage over the one who can only demonstrate traditional procurement management.

Teaching is becoming a career advantage.

The best buyers who are developing and sharing AI-augmented procurement knowledge (through teaching, writing, speaking, and community leadership) are building professional brands that create career optionality. The procurement profession needs educators who can bridge the gap between AI capability and procurement practice, and the procurement professionals who fill that role are positioning themselves for influence, visibility, and opportunity that goes well beyond their current organizational roles.

THE RISKS WE HAVEN'T DISCUSSED

I've been deliberately optimistic throughout most of this book—not because I'm naive about the risks, but because I believe the trajectory is positive for procurement teams who embrace it. But honesty requires acknowledging the risks that sit alongside the opportunities.

Over-reliance on AI is a real danger.

As AI capabilities become more impressive and more integrated into procurement workflows, the temptation to trust the AI's output without adequate human oversight will grow. The procurement pro who lets the AI make decisions rather than inform them is not more productive—they're more vulnerable. Every AI failure I've described in this book—the hallucinated benchmarks, the biased evaluations, the missed context—was a consequence of insufficient human oversight. As AI gets better, the oversight may become less frequent, but it should never disappear entirely.

Skill atrophy is a risk.

When AI handles the analytical heavy lifting, the skills that procurement professionals used to develop through that analytical work—deep familiarity with data, intuitive understanding of market patterns, the ability to spot anomalies through experience—may atrophy. The sourcing professional who has never built a spend analysis manually may lack the intuition to recognize when the AI's automated analysis is wrong. The negotiator who has never researched a supplier without AI assistance may lack the judgment to assess the quality of AI-generated intelligence. Maintaining foundational skills even as AI handles most of the execution is important—and organizations should ensure their training programs include enough manual practice to build and maintain those skills.

Concentration risk in AI providers is real.

As procurement organizations become dependent on AI platforms (for spend visibility, risk monitoring, contract intelligence, and more) they're creating a new category of supplier dependency that needs to be managed with the same discipline they apply to any critical supplier relationship. What happens if your AI platform provider is acquired, changes their pricing model, or experiences a service outage? The procurement function that has become dependent on AI-generated intelligence without maintaining the capability to function without it has created a single point of failure in its own operations.

The digital divide could widen inequality.

The democratization of AI capabilities that I've celebrated throughout this book is real—but it's not universal. Organizations that can't afford enterprise AI platforms, that lack the data infrastructure to support AI adoption, or that operate in regions or industries where AI talent is scarce may fall further behind their better-resourced competitors. For the profession as a whole, ensuring that AI's benefits are broadly accessible through affordable tools, open-source resources, community knowledge sharing, and professional development programs is an ethical imperative as well as a practical one.

These risks are real, but they're manageable. They require awareness, planning, and the kind of disciplined risk management that procurement professionals are, by training and temperament, well-equipped to provide. The risks don't change the trajectory I've described in this chapter. They shape how the trajectory unfolds for any given organization—and for any given procurement professional.

THE INVITATION

I started this book with a chicken restaurant that ran out of chicken and an airbag manufacturer whose products killed people. I told you those stories because I wanted you to understand that procurement matters—that the decisions made by the people in this profession have consequences that extend far beyond the spreadsheets and conference rooms where they're made.

I've spent the thirteen chapters that followed building the case that procurement is entering a new era—one where the foundational principles that have always governed great practice are being amplified by AI capabilities that make every procurement professional more powerful, more informed, and more capable of creating real, lasting value for their organizations.

I hope you've found the journey useful. I hope the frameworks, the workflows, the examples, and the war stories have given you practical tools that you can apply in your work. And I hope the book has convinced you of something that I believe with the full conviction of thirty-plus years in this profession: this is the most exciting time to be a procurement professional that I have ever experienced.

The tools are more powerful than they've ever been. The strategic importance of the function is higher than it's ever been. The opportunities to create value (through better intelligence, better preparation, better risk management, better supplier relationships, and better negotiation outcomes) are more abundant than they've ever been.

But the tools don't use themselves. The opportunities

don't capture themselves. The value doesn't create itself. It takes procurement pros. It takes professionals who have mastered the fundamentals, embraced the technology, and committed to the continuous learning and adaptation that this rapidly evolving landscape demands.

It takes you.

So here's my invitation.

Don't just read this book and put it on the shelf. Take one idea—just one—and apply it tomorrow. Run one negotiation through the Chapter Seven preparation framework. Build one supplier dossier using the Chapter Five workflow. Analyze one contract using the Chapter Eight approach.

The Appendix at the back of this book contains the prompts to do all three (and dozens more.) The prompts are copy-and-paste ready, organized by procurement workflow. Start with one. See the result. Let that result pull you toward the next thing.

The sourcing professionals who started six months ago already have a head start. But the beautiful thing about AI-powered capability development is that it compounds fast. The procurement pro who starts today, learns from every application, and continuously improves their approach can build significant capability in a short time.

Big Ron would barely recognize the profession today. The 1994 conference room, with its stale coffee and its ritual dance of positional bargaining, feels like ancient history. And the procurement professional of 2030 (augmented by AI capabilities we're only beginning to imagine, armed with intelligence that makes every negotiation and every sourcing decision more informed than anything we can conceive today) will look back on our current moment the way we look back on the era of paper purchase orders and filing cabinets.

As the start of something. As the moment it began to change.

Be part of that change. The profession needs you.

And frankly, it's going to be one hell of a ride.

NOW GO BUILD SOMETHING.

PART FOUR
APPENDICES & REFERENCE MATERIAL

THE TOOLKIT, CASE STUDIES, AND PLATFORM DIRECTORY REFERENCED THROUGHOUT THE BOOK.

INTRODUCTION TO THE TOOLKIT

You arrived here from Chapter Twelve's invitation to take one idea and apply it tomorrow. Welcome.

The reference material that follows (prompts, case studies, and a directory of the platforms we've discussed) is your toolkit.

Use it daily. Adapt it continuously.

The prompts are the starting point. Your expertise is the differentiator.

APPENDIX A:

THE AI PROMPT PLAYBOOK FOR PROCUREMENT—YOUR READY-TO-USE TOOLKIT

Throughout this book, I've referenced prompts and workflows for using AI in procurement. This appendix consolidates, expands, and organizes those prompts into a comprehensive toolkit that you can reference every day in your work. Think of this as the appendix you dog-ear, bookmark, and come back to repeatedly—the practical reference that turns the principles and frameworks from the rest of the book into immediate, actionable capability.

A few important notes before we dive in. First, these prompts are designed for general-purpose generative AI tools—ChatGPT, Claude, Gemini, Copilot, and their equivalents. They'll work in any of these tools, though the specific output will vary somewhat by platform. Second, these are starting points, not scripts. The most effective prompts are the ones you customize for your specific situation, category, and supplier. Use these as templates and adapt them. Third, always apply the validation practices we discussed throughout Part Two—verify facts, cross-reference data, and never cite AI-generated information in a high-stakes context without confirming it from a primary source.

How to use this appendix. The chapters of this book teach the thinking. This appendix is the toolkit. Every prompt below has appeared in or been adapted from the workflows described in Chapters Five through Nine. Keep this appendix within reach when you sit down to do the work. The body of the book is for understanding; this appendix is for doing.

SECTION ONE: SPEND ANALYSIS PROMPTS

Initial Data Assessment

"I'm uploading our procurement transaction data for the past [time period]. Before we analyze it, I need you to assess the data quality. Please identify: (1) What percentage of transactions have missing or blank supplier names? (2) What percentage have category or commodity codes assigned? (3) How many unique supplier name variations exist? (4) What percentage of transactions are coded to catch-all categories like 'miscellaneous,' 'other,' or 'general'? (5) What is the date range covered? (6) Are there any obvious data quality issues—duplicate transactions, negative amounts, dates that appear incorrect? Present your findings in a summary table with recommendations for how to address each quality issue before proceeding with analysis."

Spend Classification

"I need you to classify the following procurement transactions into our category taxonomy. Here is our taxonomy with definitions: [paste your category list with descriptions]. For each transaction, use the supplier name, transaction description, GL code, and amount to determine the most appropriate category. If you are less than 80% confident in a classification, flag it for human review. Present the results in a table showing: original data, assigned category, and confidence level (high/medium/low)."

Spend Pattern Analysis

"Based on the classified spend data we've been working with, identify the following patterns: (1) Top 20 suppliers by total spend, with year-over-year trend for each. (2) Top 15 categories by total spend, with growth rate over the period. (3) Categories with the highest number of active suppliers—this suggests fragmentation and consolidation opportunity. (4) Categories where spend has increased more than 15% year-over-year—these may need strategic sourcing attention. (5) Suppliers who appear in multiple categories—this may indicate enterprise relationship opportunities. (6) Any unusual patterns—sudden spikes, seasonal anomalies, or spending concentrations in specific time periods that warrant investigation. For each pattern identified, briefly explain why it matters from a procurement strategy perspective."

Tail Spend Analysis

"Analyze the tail spend in our data—transactions below $[threshold] individually, with suppliers where total annual spend is below $[threshold]. I need: (1) Total tail spend as a dollar amount and percentage of total procurement spend. (2) Number of tail spend suppliers versus strategic suppliers. (3) The top 10 tail spend categories by total value—these are where consolidation efforts would yield the most savings. (4) Clusters of similar purchases from multiple suppliers that could be consolidated into preferred supplier programs. (5) Your estimate of savings potential if we consolidated the largest tail spend categories into managed programs, assuming 10-20% savings through volume leverage."

Maverick Spend Detection

"I'm going to provide you with two data sets: our contracted supplier list with negotiated rates, and our actual transaction data. Compare them and identify: (1) Transactions with contracted suppliers that were processed at rates different from the contracted rate—flag any that are more than 5% above the contracted rate. (2) Transactions in categories where we have preferred suppliers but where the purchase was made with a non-preferred supplier. (3) Categories where off-contract or non-preferred spending exceeds 20% of total category spend. (4) For each finding, estimate the cost of the non-compliance—the difference between what we paid and what we would have paid under the contracted terms."

SECTION TWO: MARKET INTELLIGENCE AND SUPPLIER RESEARCH PROMPTS

Market Landscape Analysis

"I'm a senior procurement professional preparing a sourcing strategy for [category] in the [geography/region] market. Our annual spend in this category is approximately $[amount]. Give me a comprehensive market overview including: (1) The major providers by estimated market share, with their key differentiators and geographic strengths. (2) Market structure—is it concentrated among a few large players, fragmented among many small ones, or something in between? (3) Recent M&A activity that has changed the competitive landscape in the past 18 months. (4) Pricing trends over the past 12-18 months, including the key cost drivers affecting pricing. (5) Emerging trends that could disrupt current market dynamics—new technologies, new business models, new entrants, regulatory changes. (6) Any supply-demand imbalances—is this a buyer's market or a seller's market right now, and why? Focus on information that would be actionable for a procurement professional preparing a sourcing strategy, not general industry background."

Supplier Financial Deep Dive (Public Company)

"Analyze [Company Name]'s financial performance and strategic position as a potential/current supplier to my organization. Based on their most recent public filings and earnings communications, provide: (1) Revenue trajectory—growth rate over past 3 years, and current guidance. (2) Profitability—gross margin and operating margin trends, including any margin pressure or expansion commentary from management. (3) The business segment that would serve our account—its specific performance, growth rate, and strategic importance to the overall company. (4) Cash flow and balance sheet health—any concerns about liquidity, debt levels, or investment capacity. (5) Strategic priorities as articulated by management—what is leadership focused on, and how does that affect their likely approach to our relationship? (6) Competitive positioning—who are they losing to, who are they winning against, and what is their differentiation strategy? (7) Three to five insights that would be most valuable for a procurement professional preparing to negotiate with this company. Cite specific data points from their filings wherever possible."

Supplier Financial Assessment (Private Company)

"I need to assess the financial health and market position of [Company Name], a privately held company that is [one of our current suppliers / a potential new supplier] in the [category] space. Since they're privately held, we'll need to work with proxy indicators. Please research and provide: (1) Estimated company size—revenue range based on employee count (check LinkedIn), office locations, and any publicly available financial indicators. (2) Growth trajectory—are they hiring or laying off? Expanding or consolidating offices? Winning or losing visible contracts? (3) Leadership stability—any recent C-suite changes, and what might they signal? (4) Market reputation—what do reviews, references, and industry mentions suggest about their quality and reliability? (5) Competitive position—how are they positioned relative to the major players in their space? (6) Any red flags—legal actions, regulatory issues, negative press, employee complaints on review sites? (7) An overall assessment of their stability and reliability as a supplier, acknowledging the limitations of what's knowable about a private company."

Cost Driver Analysis (Should-Cost Model)

"I'm preparing to evaluate a proposed price [increase/quote] from a supplier in the [category] space. Help me build a should-cost model by analyzing the key cost components. The main cost drivers in this category are: [list them—e.g., raw materials, labor, energy, transportation, overhead]. For each cost driver: (1) What has the price/cost trajectory been over the past 12 months? Cite specific indices or data sources. (2) What is the consensus forecast for the next 6-12 months? (3) What percentage of the total product/service cost does this driver typically represent in this category? Based on this analysis, what would a reasonable overall cost change be for a product/service in this category over the past 12 months? If a supplier is proposing a [X]% increase, which specific cost drivers would need to have moved, and by how much, to justify that increase?"

Competitive Alternative Identification

"I'm currently [sole-sourced / working with a limited supply base] in the [category] space and need to identify credible alternative providers. Our requirements include: [list key requirements—geographic coverage, scale, capabilities, certifications, etc.]. Annual spend in this category is approximately $[amount]. Identify the top 5-7 alternative providers that could realistically serve our needs. For each one, provide: (1) Company overview—size, geographic presence, key clients. (2) Key differentiators—what makes them different from competitors? (3) Typical pricing model and any available pricing benchmarks. (4) Strengths and potential concerns relative to our specific requirements. (5) Any publicly available information about their win rate against incumbent providers in similar situations. (6) An assessment of the realistic effort and timeline to transition from our current supplier to this alternative, including key risks. Focus on providers who could genuinely serve our scale and requirements, not aspirational options that would require significant compromise."

SECTION THREE: NEGOTIATION PREPARATION PROMPTS

Pre-Negotiation Strategy Development

"Act as a senior procurement strategy consultant with 25 years of experience in [category] sourcing. I'm preparing for a contract [renewal/negotiation/competitive bid] with [supplier type]. Here's my situation: [describe contract value, current terms, relationship history, stakeholder priorities, and any leverage points or constraints]. Based on this context, develop a negotiation preparation brief that includes: (1) Our three strongest leverage points, with the evidence or argument to support each one. (2) The supplier's likely opening position and the three main arguments they'll use to justify it. (3) Our recommended opening position, with market data or benchmarks to support it. (4) Five creative deal structure options that could create value for both parties—for each, explain what we're offering the supplier and what we're getting in return. (5) The concessions we should be prepared to make, in what sequence, and what we should expect in return for each. (6) The two or three issues where we should hold firm regardless of supplier pushback, and why. (7) Our walk-away point—the conditions under which we should exercise our BATNA rather than accept the deal on offer."

War-Gaming: Supplier Role-Play

"Act as the [title—e.g., VP of Sales, Account Executive, Regional Director] for [type of supplier company]. You've been given the following objectives by your leadership: [describe what you know or infer about the supplier's priorities—e.g., retain the account, achieve a minimum X% price increase, get adoption of a new product, protect margin, etc.]. You know the following about the buyer's situation: [describe what the supplier likely knows about you]. I'm going to present our negotiation positions one at a time. For each one, respond as the supplier would—push back on what you'd push back on, accept what you'd accept, and make counter-proposals where appropriate. After each exchange, break character briefly and tell me: (a) what the supplier's internal reaction would be to my position, (b) what the weakest part of my argument is, and (c) what I should say differently to be more effective."

Counter-Argument Preparation

"I'm preparing to challenge a [X]% price increase proposed by our [category] supplier. Their likely arguments for the increase will include: [list expected arguments—e.g., cost inflation, market tightness, investment in new capabilities, etc.]. For each argument, provide: (1) The strongest counter-argument based on market data and industry facts. (2) A specific question I should ask that will force them to substantiate their claim with evidence rather than assertions. (3) A reframe that shifts the conversation from their cost narrative to our value narrative. Help me prepare to have this conversation in a way that is firm but respectful —challenging their positions without damaging the relationship."

Creative Options Brainstorm

"I need to generate creative deal structure options for a negotiation with a [category] supplier. Here's the context: [describe the situation—contract value, both parties' known interests and constraints, relationship history, market dynamics]. The value dimensions available to trade across include: price, payment terms, volume commitments, contract length, scope, service levels, exclusivity, innovation investment, case studies/references, and risk allocation. Generate 8-10 creative deal structures that: (1) Address at least two value dimensions each. (2) Offer something genuinely valuable to the supplier (not just asking for concessions). (3) Create net value for our organization. (4) Would be realistically acceptable to a supplier in this market position. For each option, explain the trade and assess the likelihood that the supplier would find it attractive."

SECTION FOUR: CONTRACT ANALYSIS PROMPTS

Comprehensive Contract Review

"I'm uploading a supplier contract for thorough analysis. Please review the entire document and provide a structured assessment covering: (1) COMMERCIAL TERMS:

Pricing structure, payment terms, volume commitments, price escalation provisions, most-favored-customer clauses, and any financial incentives or penalties. (2) RISK PROVISIONS: Liability limitations and caps, indemnification obligations for both parties, insurance requirements, force majeure definitions, data protection obligations, warranty provisions. (3) OPERATIONAL TERMS: Service level definitions and metrics, performance measurement methodology, change management provisions, escalation procedures, reporting requirements. (4) DURATION AND EXIT: Contract term, renewal and auto-renewal provisions (including specific trigger dates based on effective date), termination rights for both parties with notice requirements, transition assistance obligations. (5) RED FLAGS: Any provisions that are unusually aggressive, ambiguous, or that create significant one-sided risk. For each section, note whether the provision favors the buyer, the supplier, or is balanced. Highlight the five provisions that should be the highest priority for improvement in the next negotiation."

Contract Comparison

"I'm uploading [number] contracts from the same category—these are agreements with different suppliers for similar services. Please compare them on the following dimensions and present the comparison in a table format: (1) Pricing structure and effective rates. (2) Payment terms. (3) Liability caps (absolute amount and as a ratio to contract value). (4) Termination provisions and notice periods. (5) Auto-renewal terms. (6) Service level definitions and remedies. (7) Most-favored-customer provisions (present or absent). (8) Data protection provisions. After the comparison, identify: (a) which agreement has the most buyer-favorable terms overall, (b) specific provisions from each agreement that represent 'best of breed' and should be standardized across all agreements, and (c) provisions in any agreement that are materially below the standard set by the others and should be prioritized for renegotiation."

Renewal Preparation

"This contract expires in [X] days. The auto-renewal notification deadline is [Y] days. Based on your review of the current terms, prepare a renewal preparation brief that includes: (1) Key terms that should be improved in the renewal—where are we below market standard? (2) Provisions that are missing and should be added—most-favored-customer clause, benchmarking rights, enhanced termination flexibility, etc. (3) Pricing provisions—how has our pricing compared to market movement over the contract term? Is there a case for a reduction, a hold, or is an increase justified? (4) Risk provisions that need updating—have regulatory requirements changed since the contract was signed? Has the supplier's risk profile changed? (5) A prioritized list of negotiation objectives for the renewal, ranked by financial impact."

SECTION FIVE: SUPPLIER RISK ASSESSMENT PROMPTS

Comprehensive Risk Scan

"Conduct a risk assessment of [Supplier Name], one of our critical suppliers in the [category] space. This is a [public/private] company headquartered in [location]. Please assess the following risk dimensions: (1) FINANCIAL RISK: Recent financial performance, credit indicators, profitability trends, debt levels, and any signs of financial stress. (2) OPERATIONAL RISK: Any reports of quality issues, delivery failures, capacity constraints, or operational disruptions. (3) LEADERSHIP RISK: Recent executive changes, leadership stability indicators, organizational restructuring. (4) MARKET RISK: Competitive position trajectory—are they gaining or losing market share? Are there market shifts that could undermine their business model? (5) REGULATORY/COMPLIANCE RISK: Any regulatory actions, compliance violations, sanctions exposure, or pending litigation. (6) CYBERSECURITY RISK: Any reported data breaches, security incidents, or publicly visible security vulnerabilities. (7) GEOPOLITICAL RISK: Geographic concentration of operations in regions with political instability, trade restriction exposure, or natural disaster vulnerability. For each dimension, rate the current risk level (low/moderate/elevated/high) and explain your reasoning. Identify the top three risks that warrant closest monitoring or proactive mitigation."

Event-Triggered Deep Dive

"[Supplier Name], one of our key suppliers, has just experienced [describe the event—stock price drop, executive departure, regulatory investigation, natural disaster near their facility, etc.]. I need an immediate assessment. Please analyze: (1) What do we know about this event from available sources? (2) What is the likely impact on the supplier's ability to perform on our contracts? (3) What are the comparable historical situations—when other companies experienced similar events, what happened? (4) What are our contractual protections—based on the contract provisions we've discussed, what rights do we have if this situation impacts service delivery? (5) What are the three most important actions we should take in the next 48 hours? (6) What are the medium-term actions we should plan for if the situation deteriorates? Be specific and actionable—I need to brief my leadership team within 24 hours."

SECTION SIX: STAKEHOLDER COMMUNICATION PROMPTS

Executive Briefing

"Help me prepare a five-minute executive briefing on [topic—e.g., the upcoming sourcing strategy for technology services, the risk profile of our top suppliers, the results of our latest spend analysis]. The audience is [describe—e.g., CFO, executive committee, business unit president]. They care about [describe their priorities—cost, risk, speed, quality, innovation]. Structure the briefing as: (1) The headline—one sentence summarizing the key message. (2) The context—two to three sentences on why this matters now. (3) The key findings or recommendations—no more than three, each with supporting data. (4) The ask—what you need from the audience (approval, resources, alignment, decision). Keep the language crisp, avoid procurement jargon, and focus on business impact rather than procurement process."

Stakeholder Alignment Message

"I need to send a [email/message] to [stakeholder role] about [topic—e.g., an upcoming sourcing initiative that will affect their department, a supplier change that requires their input, a contract renewal where I need their priorities]. The tone should be [collaborative/informative/requesting action]. Key points to cover: [list them]. The stakeholder's likely concerns are: [list them]. Draft a message that addresses their concerns proactively, clearly states what I need from them and by when, and positions procurement as a strategic partner rather than a bureaucratic gatekeeper."

HOW TO CUSTOMIZE THESE PROMPTS: THE ADAPTATION FRAMEWORK

These prompts are templates. The best results come from adapting them to your context. Here's a framework for customization that will consistently improve your output quality.

Add organizational context.

The more the AI knows about your specific situation—your industry, your organization's size and structure, your procurement maturity level, your technology environment—the more relevant its output will be. Don't rely on generic prompts for specific situations. Tell the AI who you are, what you're trying to accomplish, and why.

Specify the output format.

If you want a table, say so. If you want an executive summary limited to one page, say so. If you want the analysis structured around questions that your leadership will ask, list those questions. The AI is remarkably responsive to format specifications—but it can't read your mind about what format you need.

Provide examples of good output.

If you have a previous analysis, briefing, or deliverable that represents the quality and format you're looking for, show it to the AI. "Here's an example of a supplier dossier I created previously. I need a similar dossier for a different supplier, matching this structure and level of detail." The AI will pattern-match to your example remarkably well.

Iterate rather than one-shot.

The first response is a starting point, not a final product. Review it, identify what's good and what's missing, and refine. "This is good, but I need more depth on the financial analysis. Specifically, I need you to compare their margin trajectory to their two main competitors and assess whether the margin pressure they're experiencing is company-specific or industry-wide." Each iteration produces progressively more useful output.

Challenge the output.

After the AI produces its analysis, ask it to critique its own work. "What are the three weakest assumptions in this analysis? What information would you need to improve the accuracy of these conclusions? What perspective might I be missing?" This self-critique step consistently surfaces gaps and biases that improve the final product.

The sourcing professional who develops fluency with these prompts (who customizes them for their categories, their suppliers, and their organizational context) is building a capability that compounds over time. Each successful application teaches you something about what works and what doesn't. Each refined prompt produces better output than the last. And the cumulative effect is a level of analytical capability and preparation depth that transforms the quality of your procurement practice.

Keep this appendix within reach.
Use it daily. Adapt it continuously.
And remember: the prompts
are the starting point. Your
expertise is the differentiator.

APPENDIX B:

CASE STUDY COLLECTION

The following case studies are drawn from real procurement situations, with details composited and anonymized to protect confidentiality. Each illustrates a specific principle or capability discussed in the main text.

CASE STUDY 1: THE INVISIBLE $14 MILLION—HOW AI SPEND ANALYSIS REVEALED A HIDDEN CATEGORY

Industry: Multi-Business B2C Enterprise | **Spend Area:** Marketing Services | **Outcome:** $14.2M in previously invisible spend identified; $2.8M in savings captured in year one

Situation:

A $6 billion multi-brand B2C enterprise believed it had strong visibility into its procurement spend. The procurement team had been conducting quarterly spend analyses for years, and the CPO was confident that all major categories were under strategic management. The organization's formal spend under management was approximately $2.1 billion across sixty-three defined categories.

When the organization deployed an AI-powered spend classification tool, they fed it three years of transaction data from all sources—including purchasing cards, expense reports, and several regional ERP instances that had previously been excluded from the quarterly analysis because integrating them was too labor-intensive.

What AI Revealed:

The AI classification identified $14.2 million in annual marketing services spend that had been coded across eleven different categories—"professional services," "creative services," "IT services," "consulting," "temporary staffing," and six others. No single category showed enough marketing services spend to trigger strategic sourcing attention. But when the AI recognized the pattern (dozens of suppliers providing substantially similar marketing and creative services, coded to different categories by different business units) the total picture was dramatic.

The fragmentation was equally revealing. Twenty-three different marketing services suppliers were being used across seven business units. Rates ranged from $85 per hour for comparable junior creative talent to $225 per hour. No master services agreement existed. No preferred supplier program governed the category. Each business unit was sourcing independently, with no leverage and no consistency.

Actions Taken:

The procurement team established marketing services as a distinct category, ran a competitive sourcing event that consolidated from twenty-three suppliers to four preferred providers, and negotiated master service agreements with volume-based rate tiers. The consolidated rates averaged thirty percent below what the highest-paying business unit had been paying and twelve percent below the organization-wide average.

Results:

$2.8 million in first-year savings from rate consolidation and volume leverage. Improved quality through a curated, vetted supplier panel. Simplified administration—four MSAs instead of dozens of ad hoc engagements. And a CPO who became the organization's most vocal advocate for AI-powered spend analytics.

Key Lesson:

The savings weren't hiding in a single dark corner. They were distributed across a dozen category codes in amounts that were individually unremarkable but collectively massive. Only AI, with its ability to recognize patterns across millions of transactions and identify clusters of similar spending coded to different categories, could have assembled the picture.

CASE STUDY 2: NINETY MINUTES THAT SAVED $1.1 MILLION—AI-POWERED NEGOTIATION PREPARATION

Industry: Financial Services | **Spend Area:** IT Infrastructure | **Outcome:** $1.1M improvement over original target on a $7.5M renewal

Situation:

A large financial services firm was approaching a renewal of its data center managed services contract—$7.5 million annually, with the incumbent supplier for six years. The supplier had proposed a seven percent price increase, citing increased security requirements and infrastructure investment. The procurement team's initial assessment was that they could negotiate the increase down to three to four percent—a "good" outcome that would satisfy leadership.

The category manager decided to apply the Chapter Seven preparation framework before accepting the "good" outcome as the target.

AI-Powered Preparation:

In approximately ninety minutes, the category manager built a comprehensive dossier. The market intelligence analysis revealed that data center managed services pricing had been declining industry-wide, by approximately four to five percent annually, as automation, cloud migration, and competitive pressure drove costs down. The supplier financial analysis (the supplier was publicly traded) revealed that their managed services division was underperforming their cloud division and that management was under pressure to grow managed services revenue through client retention rather than new business development. A contract analysis revealed that the current agreement lacked a benchmarking clause, had an auto-renewal provision triggering in forty-five days, and included a non-compete provision that had been preventing the procurement team from evaluating alternatives during the contract term.

Negotiation Outcome:

Armed with this intelligence, the category manager shifted the negotiation from "how do we minimize the increase?" to "the market says pricing should be declining, so help us understand why you're proposing an increase." The supplier's sales team was visibly surprised by the depth of market data the procurement team presented. The negotiation ultimately produced: a three percent price reduction (not a three percent increase—a net swing of ten percentage points from the supplier's opening position), removal of the non-compete provision, addition of a benchmarking clause, and a two-year term rather than the three-year term the supplier had been pushing for.

Results:

The three percent reduction on $7.5 million represented $225,000 in annual savings—$450,000 over the two-year term. But the structural improvements were worth more: the benchmarking clause alone was estimated to be worth $300,000-600,000 over time based on the organization's experience with benchmarking provisions in other contracts. Total incremental value versus the "negotiate the increase down to 3-4%" baseline: approximately $1.1 million.

Key Lesson:

The procurement team had been prepared to accept a "good" outcome, reducing the increase from seven percent to three or four percent, because their information was limited to last year's benchmarks and gut feel. AI-powered preparation revealed that the market had moved in their favor, that the supplier was more motivated to retain the business than their confident opening suggested, and that the contract contained structural weaknesses that should be addressed. The ninety minutes of AI-powered preparation didn't just save money—it fundamentally changed what the team believed was possible.

CASE STUDY 3: THE SUPPLIER NOBODY WAS WATCHING—PREDICTIVE RISK MANAGEMENT IN ACTION

Industry: Pharmaceutical | **Spend Area:** Specialty Active Ingredient Sourcing | **Outcome:** Avoided estimated $8M production disruption through early detection

Situation:

A pharmaceutical manufacturer relied on a mid-sized specialty supplier for a critical active ingredient used in two high-revenue products. The supplier was privately held, had been reliable for seven years, and was consistently rated "green" in annual risk assessments. The category manager considered the relationship stable and low-maintenance.

The organization had recently implemented an AI-powered risk monitoring platform that continuously tracked financial, operational, and reputational signals for all suppliers in the critical tier.

What AI Detected:

Over a period of approximately four months, the platform identified a converging pattern of low-level signals. The supplier's employee count on LinkedIn dropped by twelve percent. Three negative Glassdoor reviews appeared mentioning "delayed paychecks" and "uncertain future." A small trade creditor filed a lien against the supplier in the local county court. And the supplier's primary upstream supplier, a tier-two entity that the pharmaceutical company had never directly monitored, issued a press release about tightening payment terms for customers in their segment.

No single signal would have triggered an investigation. Together, the AI's pattern recognition identified the cluster as consistent with pre-failure financial distress patterns in its training data, and flagged the supplier with an elevated risk alert.

Response:

The category manager initiated a discreet financial health inquiry—framed, consistent with the Chapter Six guidance, as a partnership conversation rather than an accusation. The supplier's CEO, after some initial reluctance, acknowledged that they were experiencing cash flow challenges due to a major customer's bankruptcy and were in discussions with their bank about a credit facility restructuring.

Armed with this understanding, the procurement team executed a three-track response.

Track one: accelerated qualification of an alternative supplier, moving a relationship that had been in "exploratory" status for two years into active qualification with a target timeline of ninety days.

Track two: modified payment terms with the existing supplier—moving from net-sixty to net-thirty, reducing their cash flow pressure and buying time for the bank restructuring to complete.

Track three: built a ninety-day strategic safety stock of the critical active ingredient, enough to sustain production through a transition if the supplier's situation deteriorated further.

Results:

The supplier's financial situation stabilized over the following six months—the bank restructuring was completed, a new customer was onboarded that replaced the lost revenue, and operations normalized. The procurement team's early intervention, the accelerated payment terms and the strategic buffer stock, was credited by the supplier's CEO as having been "the difference between surviving and not." The alternative supplier was fully qualified and remained available as a backup, improving the organization's structural resilience regardless of the outcome.

The estimated cost of a supply disruption, based on the production volume at risk and the time required for emergency alternative qualification, was approximately $8 million in lost production and expedited sourcing costs. The cost of the intervention (accelerated payment terms, buffer stock, alternative qualification) was approximately $400,000. The risk monitoring platform subscription was approximately $85,000 annually.

Key Lesson:

Traditional risk management (the annual questionnaire, the green/yellow/red scorecard) would not have detected this situation until the supplier missed a delivery or announced they were closing. The AI detected the pattern months before the crisis would have materialized, creating a window for proactive intervention that prevented a disruption rather than merely responding to one.

CASE STUDY 4: THE CONTRACT PORTFOLIO THAT WAS HEMORRHAGING VALUE

Industry: Technology | **Spend Area:** Software Licensing | **Outcome:** $3.4M in recoverable value identified across 127 contracts

Situation:

A mid-cap technology company had approximately 127 active software licensing agreements representing roughly $45 million in annual spend. Contract management was decentralized—individual IT managers owned relationships with their software vendors, and procurement was involved primarily at the point of initial purchase or major renewal. The contract documents were stored across multiple systems, including the procurement platform, a SharePoint site, individual email archives, and at least one manager's personal filing system.

The newly hired CPO suspected that the portfolio contained significant unrealized value but had no way to quantify it without a comprehensive review—which, estimated manually, would take a dedicated paralegal working full-time for approximately four months.

AI-Powered Contract Intelligence:

The CPO engaged a contract intelligence initiative that used NLP to extract and analyze the key terms from all 127 agreements in approximately three weeks (including document collection, extraction, validation, and analysis).

What the Analysis Revealed:

Seventeen contracts had auto-renewal clauses triggering within the next six months—five of which had notification deadlines within the next sixty days that nobody had been tracking. Estimated value at risk from unmanaged auto-renewals: $1.2 million in pricing that could have been renegotiated.

Twenty-three contracts contained pricing that was more than fifteen percent above the benchmarks for comparable agreements in the portfolio—suggesting either that the original negotiation had been weak or that market pricing had moved favorably since the agreement was signed. Estimated addressable savings through renegotiation: $1.4 million.

Eight contracts contained benchmarking or most-favored-customer provisions that had never been exercised. Based on the portfolio benchmark analysis, exercising these provisions could recover approximately $600,000 in pricing adjustments.

Fourteen contracts contained price escalation clauses tied to CPI or other indices. In three cases, the supplier had been applying annual increases even in years when the referenced index had been flat or negative. Estimated overcharge recovery: $220,000.

Six contracts had been amended multiple times, and in two cases the cumulative amendments had created internal inconsistencies—pricing terms in one amendment that contradicted pricing terms in another. These inconsistencies created legal ambiguity that could have been exploited by either party and needed immediate remediation.

Results:

Total value identified: approximately $3.4 million—a combination of addressable savings, recoverable overcharges, and at-risk value from unmanaged renewals. The CPO prioritized the immediate-timeline items (auto-renewal deadlines and overcharge recoveries) and built a twelve-month roadmap for addressing the remaining opportunities through renewal negotiations.

Key Lesson:

The value wasn't hidden because people were negligent. It was hidden because 127 contracts, each containing dozens of provisions, stored across multiple systems, amended multiple times, and managed by different people—that volume of complexity simply cannot be managed manually with any degree of thoroughness. AI made the invisible visible, and the $3.4 million it found in a $45 million portfolio (7.6% of total spend) was remarkably consistent with industry estimates that organizations lose five to fifteen percent of contract value through poor contract management.

CASE STUDY 5: BUILDING THE AI-AUGMENTED TEAM—FROM SKEPTICS TO CHAMPIONS

Industry: Life Sciences | **Spend Area:** Full Procurement Function | **Outcome:** Team transformation over 12 months

Situation:

A specialty pharmaceutical company with a thirty-person procurement team and approximately $600 million in managed spend wanted to incorporate AI into their procurement practice. The CPO was enthusiastic, the team was skeptical, and the CIO was cautious about data security.

Approach:

Rather than launching a formal transformation program, the CPO selected three "pioneer buyers"—one enthusiastic early adopter, one respected senior leader who was mildly skeptical, and one junior analyst with strong analytical skills. Each was given an AI tool subscription, a half-day of hands-on training, and a specific assignment: prepare for their next significant negotiation using the AI-powered preparation framework, document the process and results, and present to the broader team.

Month 1-3 Results:

All three pioneers reported significant time savings (ranging from sixty to eighty percent reduction in preparation time) and measurably improved negotiation outcomes. The senior leader, whose endorsement carried the most weight with the skeptical majority, reported that her AI-powered preparation for a $4 million IT services renewal produced analysis she "never would have found manually" and contributed to an outcome that was eleven percent better than her target. She became the initiative's most vocal advocate.

Month 4-6 Rollout:

Based on the pioneers' results, the CPO offered voluntary training workshops—explicitly not mandatory. Nineteen of the remaining twenty-seven team members signed up. The training was hands-on: each participant brought a real procurement task and completed it during the workshop using AI tools, with coaching from the pioneer procurement professionals. By month six, twenty-two of thirty team members were using AI tools regularly.

Month 7-12 Integration:

The team developed shared prompt libraries, standardized workflows for common procurement tasks, and a monthly "AI practice group" where procurement teams shared discoveries and troubleshot challenges. The CPO established governance guidelines for data security and output validation. The CIO, reassured by the governance framework and impressed by the documented results, approved procurement's adoption of a purpose-built spend analytics platform.

Results at Twelve Months:

Category strategy cycle time reduced from eight weeks average to three weeks. Negotiation preparation time reduced by seventy percent. Contract compliance monitoring moved from quarterly manual reviews to continuous automated tracking. Stakeholder satisfaction scores improved significantly—business unit leaders rated procurement as "more strategic" and "more responsive" than in the prior year's survey.

Most importantly, the team's capacity expanded without headcount increases. The thirty-person team was managing more categories with greater depth, conducting more thorough negotiation preparation, and engaging more strategically with stakeholders—because AI had freed them from the data collection and report generation work that had consumed the majority of their time.

Key Lesson:

The transformation succeeded because it was pragmatic rather than programmatic. It started with three people and three assignments, not a twelve-month roadmap and a consulting engagement. It scaled through demonstrated results and voluntary adoption, not mandates and compliance tracking. And it was led by procurement professionals, including a skeptic who became a champion, rather than by a technology team that didn't understand procurement's daily reality.

APPENDIX C:

THE PROCUREMENT AI TOOLKIT —A PRACTITIONER'S DIRECTORY

This directory provides an overview of the AI tools and platforms referenced throughout this book, organized by procurement function. The landscape is evolving rapidly—tools are being updated, new entrants are appearing, and capabilities are expanding continuously. Use this as a starting point for your evaluation, not as a definitive ranking.

GENERAL-PURPOSE AI TOOLS (LAYER TWO)

These tools are not procurement-specific but are extraordinarily useful for procurement work when directed by a procurement professional who knows what to ask.

ChatGPT (OpenAI)—The most widely used generative AI tool. Strong for market research, document analysis, communication drafting, data analysis, and negotiation preparation. Available in free (limited) and paid ($20/month) tiers. The paid tier includes advanced data analysis, file upload, and image capabilities. Best for: broad research, drafting, and analysis tasks.

Claude (Anthropic)—Particularly strong at analyzing long documents, including contracts and RFP responses. Handles nuanced analytical tasks well and tends to be more cautious about stating uncertain information as fact. Best for: contract analysis, detailed document review, and complex analytical tasks.

Gemini (Google)—Integrated with Google's search capabilities, making it strong for research tasks that benefit from current web information. Best for: market research and supplier intelligence where current information matters.

Microsoft Copilot—Integrated into the Microsoft 365 suite, allowing AI-assisted work directly within Excel, Word, PowerPoint, and Outlook. Particularly useful for procurement pros whose workflow is centered on Microsoft tools. Best for: data analysis in Excel, presentation creation, and email communication.

PROCUREMENT-SPECIFIC AI PLATFORMS (LAYER THREE)

Spend Analytics

Coupa—Comprehensive business spend management platform with AI-powered spend classification, community benchmarking data from thousands of organizations, and predictive analytics. One of the largest community data sets in procurement, enabling peer benchmarking at scale.

SAP Ariba—Enterprise source-to-pay platform with embedded AI through the Joule copilot. Strong integration with SAP ERP environments. AI capabilities include category management recommendations and natural language search across procurement data.

Suplari—Focused spend intelligence platform providing AI-powered spend classification, supplier normalization, and actionable insights. Known for surfacing savings opportunities and compliance gaps through automated pattern detection.

Supplier Risk Monitoring

Resilinc—Market-leading supply chain risk management platform. EventWatch AI monitors over 100 million data sources worldwide for potential disruptions. Provides multi-tier supply chain mapping, predictive risk modeling, and automated alert and response capabilities. Named a Leader in Gartner's 2025 Magic Quadrant for supplier risk management solutions.

QIMA Signals—AI-powered risk monitoring using the Kavida.ai engine. Provides early warning of geopolitical, natural disaster, epidemic, and ESG risks through continuous monitoring of global data sources.

Contract Intelligence

NLP-powered platforms that read, extract, and analyze contract terms at scale. Key capabilities include automated term extraction, portfolio benchmarking, renewal tracking, obligation management, and risk pattern detection across the contract portfolio.

Intake and Orchestration

Omnea—AI-native intake and orchestration platform that transforms how procurement requests enter the system. Uses AI to auto-populate request details, flag policy divergence, generate RFx questionnaires, and provide contextual summaries for approvers.

Tonkean—No-code orchestration platform with ProcurementGPT capability that auto-generates RFPs, compares vendor options, and creates compliant purchase orders—particularly useful for tail spend and routine purchases.

Autonomous Sourcing

Procure AI—AI-native procurement automation platform with 40+ AI agents covering the full procurement process. Capabilities include autonomous tactical sourcing and negotiation, spend analytics, supplier management, and intake management. Published results include 37% shorter order processing times, 47% faster award decisions, and approximately 5% savings on tail spend negotiations.

Procurement pro-Built Platforms

ProcureOptima—Strategic procurement platform designed by procurement professionals for procurement professionals, focused on sourcing productivity and strategic decision support. Represents the emerging category of tools built by sourcing professionals with dual expertise in procurement and technology, designed around the workflows that procurement teams actually use.

Source-to-Pay Suites with AI Capabilities

Zycus—Merlin GenAI Suite integrated with Microsoft Teams for conversational AI assistance across procurement tasks. Pioneer in generative AI application to source-to-pay processes.

Jaggaer—AI-powered spend analytics, supplier management, and e-procurement with a focus on strategic sourcing and supplier performance optimization.

GEP—Unified source-to-pay platform with AI capabilities across spend analysis, sourcing optimization, contract management, and supplier risk assessment.

Ivalua—Highly configurable procurement platform with AI capabilities that can be tailored to organization-specific needs, strong integration capabilities, and a focus on user configurability.

EVALUATING PROCUREMENT AI TOOLS: A QUICK REFERENCE FRAMEWORK

When evaluating any procurement AI tool, assess it against these five criteria:

1. **Problem-Solution Fit:** Does it solve a problem that is genuinely painful for your organization, or is it a solution looking for a problem?
2. **Data Integration:** Can it connect to your existing systems (ERP, S2P, contract repository) and process your actual data, or does it require extensive data preparation?
3. **AI Transparency:** Can you understand why the

tool produced a particular result? Is the reasoning explainable and auditable?

4. **Procurement Expertise:** Was the tool built by people who understand procurement, or by technologists who think they understand procurement? Talk to the implementation team, not just the sales team.
5. **Total Cost of Ownership:** What does the tool cost over three years, including subscription, implementation, training, integration, and ongoing support? And does the business case—grounded in your quick win results—justify that investment?

Note: This directory reflects the landscape as of the date of publication. The procurement AI market is evolving rapidly. For the most current information on tools and platforms, visit my website for updated reviews and assessments.

dcjconsultinggroup.com

www.ingramcontent.com/pod-product-compliance
Lightning Source LLC
LaVergne TN
LVHW010710061126
851857LV00014B/1122